The Archetype of the Number
and its Reflections in
Contemporary Cosmology

© 2018 by Chiron Publications. All rights reserved. No part of this publication may be reproduced, stored in a retrieval system, or transmitted, in any form by any means, electronic, mechanical, photocopying, recording, or otherwise, without the prior written permission of the publisher, Chiron Publications, P.O. Box 19690, Asheville, N.C. 28815-1690.

www.ChironPublications.com

Interior and cover design by Cornelia G. Murariu
Printed primarily in the United States of America.

Front Cover Image: *Holoscope* 2011, Simeon Nelson, CAD drawing
Interior figures, unless otherwise noted, are by author or public domain images.

ISBN 978-1-63051-438-9 paperback
ISBN 978-1-63051-439-6 hardcover
ISBN 978-1-63051-440-2 electronic
ISBN 978-1-63051-587-4 limited edition paperback

Library of Congress Cataloging-in-Publication Data Pending

The Archetype of the Number and its Reflections in Contemporary Cosmology

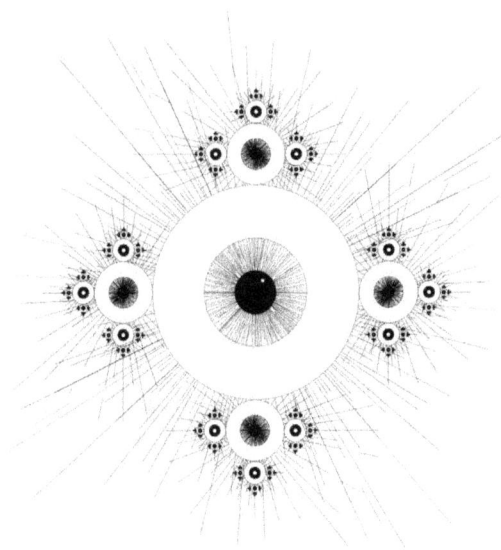

ALAIN NEGRE

"During those years, between 1918 and 1920, I began to understand that the goal of psychic development is the self. There is no linear evolution; there is only a circumambulation of the self."[1]

CARL GUSTAV JUNG

"... the quaternity is the sine qua non of divine birth and consequently of the inner life of the trinity. Thus, circle and quaternity on one side and the threefold rhythm on the other interpenetrate so that each is contained in the other."[2]

CARL GUSTAV JUNG

"But the moment when physics touches on the 'untrodden, untreadable regions,' and when psychology has at the same time to admit there are other forms of psychic life besides the acquisitions of personal consciousness – in other words, when psychology too touches on an impenetrable darkness – then the intermediate realm of subtle bodies comes to life again, and the physical and the psychic are once more blended in an indissoluble unity. We have come very near to this turning-point today."[3]

CARL GUSTAV JUNG

"All things do live in the three/ But in the four they merry be."(Squaring the circle)[4]

(Reprinted with permission from PUP and Taylor & Francis)

---

1 C.G. Jung, *Memories, Dreams, Reflections,* (New York, NY: Pantheon Books, 1963), 196.
2 Jung, *Psychology and Religion: West and East,* CW 11, 2nd ed., trans. R.F.C. Hull (Princeton, NJ: Princeton University Press, 1969 and London, UK: Taylor & Francis Ltd, 1970), §125, p. 73.
3 Jung, *Psychology and Alchemy,* CW 12, tr. R.F.C. Hull (Princeton, NJ: Princeton University Press, 1980 and London, UK: Taylor & Francis Ltd, 1989), § 394.
4 Herbrandt Jamsthaler, Viatorium spagyricum. *Das ist: Ein gebenedeyter spagyrischer Wegweiser,* (Frankfurt am Main, 1625), 272. In Jung, Psychology and Alchemy, § 165.

Table of Contents

Preface

As far as I can remember, I have always been curious about the complex relationships between consciousness, mind, the brain, and the external world. From the fluctuating journey towards this book, I recognize three milestones like so many distant and foggy memories (which could very well have been four).

The first is the sense of wonder and mystery I had in my discovery of physics at the Lycée of Moissac, a small town in the southwest of France. Mechanics, but especially electromagnetism and induction phenomena made me question my whole being and struck my teenage imagination. It was, above all, the surprising agreement between experiment and mathematical relation – an intense but very short amazement – that dissolved within a few seconds at most during the first lab. I learned much later that this astonishing effectiveness of mathematics in accounting for nature's laws – so effective that it seems to go without saying – is still not understood today. At that time, when the mellifluous melody of "Strawberry Fields Forever" was in the air, it seemed to me that this feeling of astonishment which had seized me was not that different from what I experienced some time later in discovering that certain elements of my inner life were reflected in my chart – although incomplete for being limited to the sun and rising signs. John Lennon's captivating voice soared then, over the dreamy sounds of the mellotron and the swarmandal, blended with the otherworldly vagueness of the lyrics: "[…] Nothing is real, and nothing to get hung about, strawberry fields forever […] Always, no sometimes, think it's me, but you know I know when it's a dream, I think I know, I mean: Ah, yes, but it's all wrong, that is I think I disagree…"

The second milestone was not so far from the date of the death of the creator of 'Imagine'. Driving my car in the middle of Paris – just this once – my attention was so drawn to the radio that I had to stop and park the vehicle. It was the retransmissions of the famous international Symposium of Cordoba, "Science and Consciousness: Two Views of the Universe." Researchers from around the world had gathered to discuss science and spirituality, a first, which sparked some controversy related to what some saw as an intolerable "mixing of genres". At about that same time, I began reading Carl Gustav Jung and his continuator Marie-Louise von Franz, particularly on the topic of the number as a common ordering factor of the psyche and matter. I was surprised – but also comforted – to learn that the physicist Wolfgang Pauli, known for his work on atoms, was not only interested in the same psychophysical issues, but had actively collaborated with Jung for a quarter of a century. In the wake of this, the stimulating reading of "Science and the World Soul" by Michel Cazenave, organizer of Cordoba, showed the possibility of a point of unity in the multiplicity of disciplines in an essential dialectic of the knowledge of the unconscious and rational processes – the central idea being that of the essential complementarity of science and mysticism in a global reunification which respects the specificity of each of these domains.

Finally, in the 2010s my attention was drawn to some theoretical advances in cosmology that pointed to the same 'polarity' that I had brought to light in a first book published in 1994. In it, I had stood up against the thesis of some astrologers who regarded their art as a science. The third chapter was entirely dedicated to a parallel between the history of the universe and the symbolic structure of the 12 signs of the zodiac. Therefore, while taking advantage of Pauli's work (finally!) translated into French, the idea took shape in my mind to explore the same parallel, while choosing the number as a thread. Wondering about the role of the latter as an intermediary between Being and its manifestations, I wrote this new book and translated it into English shortly after in order to reach a wider audience.

Saint-Martin d'Hères, March 2018

1. Seeing the History of the Universe in a Different Light

The universe as a story and a temporary structure.
Numbers and narratives. The subtle reality of the
number.

The first structurings of the world were the constellations: groups of bright stars connected by imaginary lines reproducing human or animal images. These remarkable forms of a sidereal mythology are not flat, as they result from a perspective effect in a space now understood as three-dimensional. In fact, advances in astronomy have awakened consciousness of the sky's depth, and recovered from the veil of oblivion stories of sidereal deities. Thus, today, the ancient cosmologies have been virtually eliminated and appear to offer nothing more than historical interest, as with astrology and alchemy.

Today, when we observe the universe, we realize that it remains under construction because to our eyes, and through the observational and conceptual tools that extend them, the universe appears as a temporary structure. Indeed, the light arriving from distant objects – sometimes separated by several billion light-years – shows these stars as they were billions of years ago.

Consequently, the predominant model of the universe is interpreted as a history whose theories explain a primordial phase of differentiation between matter and light. Today we observe in the deep background of the sky a pale reflection of this event which occurred when the universe was 380,000 years old.

Then a slow and gradual structuring of matter in galaxies and stars occurred, and the latter created heavy elements. Our planet was formed by accretion 4 billion years ago. Then, 3.8 billion years ago, life came in the form of molecules that replicated. Finally, animal and human species appeared.

Modern physics can extrapolate future events of the universe, but despite the extraordinary effectiveness of the *Big Bang* model, its history is dotted with unexplained discontinuities and transitions, with gaps which are more or less obscure. Beyond this story, of which the main steps are known, considerable puzzles remain: dark matter, dark energy, the cosmological constant, etc.

And to these questions of *how*, which science can at least partially answer, are now added metaphysical questions expressed in terms of *why* and *for what?* What is the meaning of this strange complicity between the laws of physics and our presence in the universe?

Obviously, the Cartesian restrictive framework that was very useful to science is no longer suitable. The problems of the relationship between mind and body, often described in terms of a circular process, oppose the modern notion of causality. In this view, evolutionary theory – having become an intrinsic part of cosmology – requires us to reformulate the questions dealing with the role of the observer in the object studied.

Reformulating the problem, or seeing it differently in order to surpass it, is what the scientists behind quantum theory had done in the early decades of the twentieth century after realizing that the classical concepts were no longer adequate. To this end, Niels Bohr and his colleagues turned to certain concepts of the philosophers of antiquity, or found inspiration through thinking in ways foreign to their culture.

Science advances in a cumulative manner as time moves forward, but we must not forget that it is based on the most fundamental philosophical or religious questions. Religion is, indeed, not only about the social bond, as its usual association with the Latinate word *religare* (to bind) tends to suggest.

The most fundamental root *relegere* 'go through again', from *re* 'again' + *legere* 'read' associates it with a meditative approach of scrupulous and meticulous attention, which approximates it with rational thinking[1]. Thus, confronting modern science with traditional religious or philosophical views should not scare off the modern physicist (naturally, on the condition that one does not regress into a confused concordism ultimately benefitting no one.)

To see the problem of the historic narrative of the universe from another viewpoint, for example, could mean to reconsider the idea of causality – a central notion in physics – which has become a narrow-minded concept still associated with the idea that cause precedes effect. A return to the Aristotelian conception of causality could contribute to a new perspective on the processing of knowledge.

For Aristotle, to know something means knowing the causes related to its mode of being. Through movement, an effect of an imperfection, *being* seeks to realize, or actualize, its potential. Thus, any movement must have a cause. Aristotle uses the three known causes in his time and adds a fourth: in his eyes, the most important one.

Causality is a complex metaphysical notion that we can understand with the help of four questions concerning an object:

1. What is this object composed of? This is the combination of all parts concerned in the advent of the being by which one can say what it is. This is the *material cause*.

2. What is its form or the pattern that it exemplifies? This is the *formal cause*, or essence of the object.

3. Where does this object come from? What is the principle or movement that gave rise to it? If it is a statue, is it the sculptor of the sculpture or the art of sculpture? Later, in medieval philosophy, this *motor cause* will become *causa efficiens*. It will then become not only an explanation of the change in things, but an explanation for their existence. Things are what they are because their existence comes

from outside themselves, and what gives existence will be called the *efficient cause.*

To these three causes, Aristotle adds a fourth and final one:

4. For what purpose has the thing been made? Why has it been made? The *final cause* is the goal intended by any transformation, what it was as potential being.

At the advent of classical science, Descartes and Galileo will eliminate three of the four forms of causality used by Aristotle and only keep the motor or efficient cause. The events occurring in nature will be considered according to a strictly linear pattern: a billiard ball knocking another ball that knocks another one that strikes a third one. But today, these concepts are no longer adapted to the paradoxes that arise in the current modeling of the universe.

To return to an enriched notion of causality is just one example, among others, allowing one to see the story of the history of the universe differently. In this story, opposites seem to play a fundamental role, but are understood with difficulty by contemporary science. The modern notion of feedback loop, now widely used in physics, could potentially be enriched through a comparison with the thoughts of Heraclitus of Ephesus.

In the sixth century BC, Heraclitus observed that nature is in a state of constant flux, a view he famously expressed by saying: "You can not step twice in the same river." But while the changes are continuous, the cosmic *Logos* (a binding force) maintains an orderly balance in the world. Heraclitus saw in the apparent unity of things a multiplicity of opposite qualities. He applied the contradiction not only to simultaneous elements, but to successive ones. His concept of enantiodromia[2], literally 'running in the opposite direction' means that over time everything tends to redirect and return to itself.

Searching for new conceptual tools, it might be worth returning to an expanded concept of number. The quantitative number of contemporary mathematics is a reduction in the traditional conception of number as it appeared with Pythagoras and Plato who, in his dialogue, *Timaeus*, uses numbers to report the composition of the World Soul out of certain intermediate kinds of Existence, Sameness and Difference. But is he dealing with the same concept of number that is used today by contemporary science?

Before going further with this distinction, let us note the quasi-general irruption of stories and narratives in contemporary physics. Now that stones, Earth, the solar system, and on up to the universe as a whole are dated, interpretations of their theoretical models present themselves as narratives.

And although the number used in these mathematical theoretical models is no longer that of Pythagoras or Plato, the fact remains that a link between number and narrative has been perpetuated in universal memory. The etymology reflects the common roots of *count* and *recount* and the noun and verb *account* partakes of both arithmetic and narrative sense. As for *to tell*, the old English *tællan or tellan*, meant "To mention or name a series of things one after another in order; to enumerate."[3]

These words reflect a time when calculation and narrative did not contradict one another. Then with the evolution of ideas, the verb '*to tell*' eventually took on the meaning of enumerating facts, relating the events of a story, while the purely enumerating sense of arithmetic was left to the verb '*to count.*'

The great physicist of the twentieth century, Wolfgang Pauli, was conscious of a subtle reality, far richer than the number used in a purely scientific capacity. This qualitative aspect of number, which traverses people and things with its universal rhythm, survived the successive cuts and reductions of science, a science that favors the collective structural aspects of numbers, which are their purely quantitative aspects.

ENDNOTES

1   Francis E. Abbot, "A Study of Religion: The Name and the Thing [5th Sunday Afternoon Lectures for 1872, Horticulture Hall Boston]," in *The Pamphlet Collection of Sir Robert Stout: Volume 37, The Derivation from 'Relegere'* (2016). Available online at http://nzetc.victoria.ac.nz/tm/scholarly/tei-Stout37-t15-body-d2-d3.html
2   See Aras, The Archive for Research in Archetypal Symbolism *"Enantiodromia."* Available online at https://aras.org/concordance/content/enantiodromia
3   See Wiktionary, the Free Dictionary *"Tell."* Available online at https://en.wiktionary.org/wiki/tell

2. Wolfgang Pauli and Number as 'Primitive Mathematical Intuition'

The borderland between physics and psychology.
Number as the most basic element of order in the
human mind. The imagination as a power of the
soul. Scientific theories and independent reality.

Wolfgang Pauli, the Nobel laureate physicist of 1945, is famous for his 'exclusion principle' which says that two electrons cannot have the same set of quantum numbers. But he is also recognized for his boldness in predicting the existence of a new particle discovered afterwards, called a 'neutrino.'[4] Furthermore, the best known of his numerous works relate to considerations on symmetries.

Searching for symmetry is like looking for landmarks. Using symmetry as a guide, Pauli has largely contributed to bringing order to the increasingly complex world of elementary particles, as evidenced by a theorem he demonstrated in 1955 still used today under the name of the 'CPT theorem.' This theorem states that quantum field theory is invariant through simultaneous time reversal (T symmetry), charge conjugation (C symmetry), and space (P symmetry): the letter P comes from *parity* which, "in connection with an integer, denotes the distinction of even and odd. The application of this concept to spatial reflections arises from the circumstance that in the case of invariance of all interactions under spatial reflections, the eigenstates, according to wave mechanics, are divided into 'even' and 'odd' in such a way that the wave functions of the even states remain unchanged when the signs of all the spatial coordinates are changed (reflection), while the wave functions of the odd states change sign.

The sign thus defined, +1 for even and -1 for odd states is called the *parity* of the state."[5]

Pauli was particularly fascinated by *mirror symmetry* which, by its inevitable consonance with inside/outside, would evoke in him the conviction of deep links between mind and matter. He did not publish extensively in this area, preferring to communicate ideas by letters to colleagues and especially to the psychologist Carl Gustav Jung (Fig. 1). With a transdisciplinary spirit way ahead of his time, Pauli rigorously respected the different ways of thinking attached to each discipline.

Figure 1. Jung (1875-1961) and Pauli (1900-1958).

Reflecting on problems arising from the confrontation of the realms of mind and matter, the two scholars were led to the notion of a field of symmetry where the mind is material and matter spiritual. This realm is reminiscent of the *unus mundus*, or 'one world' of medieval philosophers, where what is separated in the manifested world is united on another level of reality, specifically that of the *unus mundus*.[6] Thus, physics and depth psychology would both lead, in their own ways, to this other level of reality.

The two researchers were early in their intuition of the importance of numbers in creating a common language between matter and psyche. Already, in the years before the development of the new quantum mechanics, Arnold Sommerfeld, a professor of the young Pauli at the University of Munich, had adopted an intuitive and symbolic approach to physics.

Shaken by Niels Bohr's revolutionary hypothesis of the atom planetary model (1913), Sommerfeld tried to interpret the empirical laws of atomic spectra using number sequences. Driven by a deep sense of harmony and fascinated by whole numbers, he permeated the minds of his students with a certain 'cabalistic' numerology. His attitude had raised the following paraphrase from a well-known advertisement of the time: "if it's about integers, go see Sommerfeld."[7]

Later, when the meaning of the electron's fourth degree of freedom – its spin – was in question, Pauli was the first to doubt the planetary analogy, suggesting that the electron, in addition to its rotation around the central core, would spin around like a top. Quantum numbers seemed much more real to him than orbits. This reality was confirmed by subsequent developments in quantum theory.

The lengthy published correspondence between Jung and Pauli shows that they mutually inspired one another on the subject of number. Pauli sought to transcend the limits of mathematical physics. He considered number as a "primary probability"[8] or "primary possibility." In doing so, he agreed with Jung's approach, for whom number was the "most primitive element of order in the human mind."[9]

The quantitative aspect of number is pure number, the collection and repetition of unity. It is what has remained of the original aspect of number after the immense effort of abstraction that has been done through the development of mathematics. Alas, the qualitative aspect of number survives today *only* through regressive activities of numerologists and seers.

However, it was through a step beyond the limits of reason that Jung and Pauli considered the qualitative aspect of number and wished to use it as a basis of representation of all physical and mental processes. They were contemporaries of the advent of quantum physics, which has given rise to paradoxes and conflicting pairs such as the wave and the particle.

Following their discussions on the links between the psyche and matter, they managed to lay down the foundations for an agreement to unify scientific and unscientific ways of thinking in the same philosophical vision. While exploring analogies between the quantum world and the psyche, they concluded that mind and matter were complementary aspects of a deeper level of reality, just as waves and particles were complementary aspects of the quantum reality.

Following this transdisciplinary perspective and building on their work, we will attempt an incursion through the maze of recent scientific advances. Guided by their assumptions on the reality of number, we will explore the scientific model of the universe, which is the provisional culmination of contemporary science.

Well before approaching the universe as a scientific object, classical physics had limited its field by bracketing the study of the human, which had made it very effective. In order to be able to describe reality, it had simplified it and reduced it to a small number of general ideas and assumptions defining space, time, and matter.

Modern physics has replaced these objects with localized entities (particles) and extended entities (fields); and today, the concept of information – already represented in classical physics by its inverse, entropy – has made its entrance. Physical laws are written in mathematical language and express general causal links, links between abstract concepts in the form of figures, geometrical diagrams, or analytical equations including physical constants (the speed of light, Planck's constant, etc.).

However, mathematics is not a simple tool that one uses in order to 'do' physics. In fact, physics has become inseparable from its mathematical form, an event that could be dated as far back as the 'Law of periods,' after the work of Johannes Kepler (Fig. 2). This law, published in 1618, allows us to determine the distance of a celestial body from the Sun provided that we know its period of revolution.

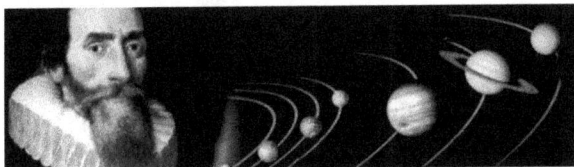

Figure 2. Johannes Kepler (1571-1630).
(Reprinted with permission from Hervé Le Cornec).

From that time on, progress in physics has almost exclusively depended upon advancements in mathematics. Today, mathematical concepts have become exactly like those in physics. As Freeman Dyson writes: "For a physicist, mathematics is not just a tool by means of which phenomena can be calculated, it is the main source of concepts and principles by means of which new theories can be created."[10] The development of physics fits within the mathematization process that closely associates mathematics and physics, making the discourse of physics consubstantial with its mathematical form.

Mathematical physics uses figures, numbers, and algebraic symbols which are all pure constructions of the human mind. But there is a good reason to be intrigued by their efficiency: how can these abstract symbols, articulated by sets of specific rules produced by the mind, account for the phenomena of the empirical world?

Pauli had asked himself these questions and had supported the side of one of his sixteenth century predecessors, Johannes Kepler, to whom he devoted an essay in 1952.[11] Kepler, a transitional character and astronomer-astrologer born in 1571, claimed the advantage of the heliocentric model. He was convinced that God had created the universe in accordance with a mathematical plan. Pauli saw that his scientific thoughts were intertwined with symbolic representations: the world was organized according to the divine trinitarian image, and the Platonic polyhedra were reinstated.

From this symbolic background, Kepler had extracted physical concepts still very underdeveloped but that had led to his famous 'three laws' that inaugurated the new science. A few decades later, Newton resumed and superseded the works of Kepler and Galileo while building, as far as he was concerned, on the symbolic images of alchemy to establish the universal law of attraction.[12]

This is what 19th century historians of science did not want to acknowledge: mathematical formulas do not arise as such from pure spirit. In fact, a long development work – based on symbolic, religious, and metaphysical representations, or else dreams – precedes rational ordering. Knowledge is thus based on the *imagination*, a term which must be distinguished from all that the Western tradition has classified as illusory and unreal.

Pauli was struck by the insistent reference that Kepler was making to the fifth century Neoplatonist, Proclus (Fig. 3), for whom *to know* is "[...] to compare that which is externally perceived with pre-existent inner ideas."[13] The latter pre-exist in the soul, which is a notion that the Neoplatonists connected to the World Soul.

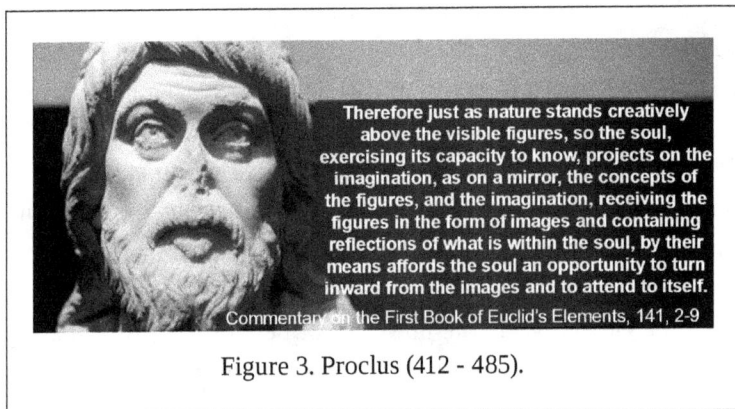

Therefore just as nature stands creatively above the visible figures, so the soul, exercising its capacity to know, projects on the imagination, as on a mirror, the concepts of the figures, and the imagination, receiving the figures in the form of images and containing reflections of what is within the soul, by their means affords the soul an opportunity to turn inward from the images and to attend to itself.

Commentary on the First Book of Euclid's Elements, 141, 2-9

Figure 3. Proclus (412 - 485).

For Proclus, the soul, or "middle and center of all beings," is capable of grasping intellectual Forms. It contains within itself all the figures and images that may prove necessary for the understanding of the sensory world. It is formed of reason-principles (*logoi*), of mathematical, logical, and ethical patterns that enable it to understand reality and that bridge the sensory and the intelligible. Following Plato and against Aristotle, Proclus thus conceived of mathematics as a power of the soul ensuring the link between spirit and matter.

In *"Mathematics and Soul in Proclus"*[14] Michel Cazenave demonstrates the parallelism between the soul and mathematics, and the paradox of the latter as being both the language of the soul and the expression of the soul as structuration and production: "[…] in the same way that soul creates a world in which it moves and deploys its capacity of knowledge, any mathematical assertion is, simultaneously, structure and knowledge of objects that fall within that structure. To make an analogy with modern science, the paradox of contemporary cosmology should be remembered: how does the universe spread into the space it creates itself? Indeed, perhaps these are only apparent paradoxes, pure language effects, and a solution may be found in the necessary procession of the productive instance in the world it has precisely produced?"[15]

Proclus brings the power of *imagination,* which receives, retains and transforms the data of the sensory world while being at the same time memory and creation. During the genesis of a geometrical form, for example, in the soul of a mathematician, the imagination "[...] is found in a median position, in a mixture in some way, between the contemplation of the pure intelligible figure and the recognition of the form that is perceived in the sensory world. In its ability to ensure the fact that mathematics, from this point of view, reveals the power of principles, they are in some way its implementation; they allow for its unfolding – which also means the dispersion – in the same movement with which they structure it: the veil of the intelligible at the same time as the permanence of the intelligible in things."[16]

In spite of the soul being put in parentheses by Newton (who replaced it by the notion of force – to be later itself replaced by the notion of field), mathematics – without publicly admitting it to itself, Cazenave continues – express the divine. It arises in a movement of procession (for example when the mathematician draws a figure on paper and begins to structure the real) and is caught up in the movement of conversion when the whole of Being turns back to its primeval origin. But science erases the traces of its own steps, and Cazenave wonders, "[...] what metaphysical phantom, if not frankly theological phantom, haunts it in its tireless pursuit of unification which is its own, which seeks to subsume the plurality of beings under the assertion of an intelligible One – of which the current program of grand unification is the latest and most contemporary symptom."[17]

If the soul, while ensuring the unity of the universe between the One and matter with all hierarchical degrees of intermediate planes, generates mathematical forms and ratio, then there would be no surprise to discover "a nature written in mathematical language."[18] Two decades after Kepler's third law, Galileo, author of this famous remark, was not surprised. In fact, he does not hide his admiration for Plato and Pythagoras: "I know very well how much the Pythagoreans held the science of numbers in the greatest esteem and that even Plato admired human intellect and considered it a participant in divinity only because human beings understand the nature of numbers. I myself am not too far from formulating the same judgement."[19] However, things went very fast: although Kepler fully availed himself of Proclus and the Neoplatonists, Galileo made a decisive step towards Aristotle instead. Hence his interest in experimentation and mechanism, which rapidly accelerated the collapse of the mediating instance of soul. Today, the astonishment at the "extraordinary effectiveness of mathematics" has become amazement before the contemporary 'miracles' of quantum physics, models experimentally verified up to the eighth decimal. This effectiveness even extends to bioinformatics and the biology of populations.

This surprise is commensurate with the oblivion of what occurred when Cartesian dualism triggered the formation process of the new science: a radical discontinuity between the world 'outside' and the world 'inside;' a split which led to the destruction of the tripartite Neoplatonic schema of the mind, soul, and body. The intermediate domain of the soul had disappeared following the development of scientific ideas. A philosophy of mind-matter dualism rose instead, and science gradually was extracted from culture. It became autonomous up to the point of alienation. How can we believe any longer, laments Jean-Marc Lévy-Leblond "[...] that science would be different in this respect than art or philosophy or literature, where it would not occur to anyone to teach these fields of human activity regardless of their history?"[20]

Inspired by Proclus and Kepler in 1952, Pauli wanted to recover this forgotten history: "The process of understanding nature as well as the happiness that man feels in understanding, that is, in the conscious realization of new knowledge, seems thus to be based on a correspondence, a 'matching' of inner images pre-existent in the human *psyche* with external objects and their behavior."[21]

By speaking of images produced by the soul, Pauli was able to overcome the traditional Western meaning that has always tended to define the imaginary as illusory, a source of error and falsehood. On the contrary, he understood imagination as a power of the soul where inventions are rooted. In this perspective, scientific knowledge is a historical process of model building; in other words, *a priori* formulations of hypotheses, invented, imagined, and designed to be closer to – without ever reaching – independent reality.

ENDNOTES

4   Neutrinos are subatomic particles that interact very rarely with matter. Antineutrinos from distant reactors simulate the disappearance of solar neutrinos. The photo on the back cover shows one of eight Daya Bay detectors situated about 32 miles northeast of Hong Kong. The detector observes correlations between reactor core fuel evolution and changes in the reactor antineutrino flux and energy spectrum. The detector's 3.1 m diameter cylindrical antineutrino target contains 20 tons of gadolinium-doped liquid scintillator and is surrounded by an array of photomultiplier tubes and shielding. In an inverse beta decay, an electron-antineutrino is detected when it interacts with a proton to produce a positron and a neutron.

5   Wolfgang Pauli, "The Violation of Reflection Symmetries in the Laws of Atomic Physics," in *Writings on Physics and Philosophy*, ed. C. Enz and K. von Meyenn (Berlin Heidelberg: Springer-Verlag, 1994), 186.

6   The Jung-Pauli approach on the so-called 'psychophysical problem' is a variant of dual-aspect monism, which has become interesting for physicists such as David Bohm and Bernard d'Espagnat. See Harald Atmanspacher, "20th Century Versions of Dual-Aspect Thinking," *Mind & Matter* Vol.12(2), (2014). http://www.mindmatter.de/journal/abstracts/mmabstracts12_2.html#atm

7   Pauli, "Sommerfeld's Contributions to Quantum Theory," in *Writings*, 65.

8   To be developed later: Pauli, "Ideas of the Unconscious from the Standpoint of Natural Science and Epistemology," in *Writings*, 164.

9   C. G. Jung, *Synchronicity: An Acausal Connecting Principle, CW 8*, 2nd ed. (Princeton, NJ: Princeton University Press, 1973), § 870.

10   Freeman Dyson, "Mathematics in the physical sciences," *Scientific American*, Vol. 211, Issue 3, September 1, (1964): 129-46.

11   Pauli, "The Influence of Archetypal Ideas on the Scientific Theories of Kepler," in *Writings*, 219-79.

12   Betty Jo Teeter Dobbs, *The Janus Faces of Genius: The Role of Alchemy in Newton's Thought* (Cambridge, UK: Cambridge University Press, 2002).

13 Pauli, ibid., 226.

14 Michel Cazenave, "Les mathématiques et l'âme chez Proclus [Mathematics and Soul in Proclus]," in Cazenave et al., *De la science à la philosophie: y a-t-il une unité de la connaissance ?* (Paris: Albin Michel, 2005), 422-48.

15 Ibid.

16 Ibid.

17 Ibid.

18 Galileo wrote to Fortunio Liceti in January 1641: "The book of philosophy [...] stands perpetually open before our eyes, though since it is written in characters different from those of our alphabet it cannot be read by everyone; and the characters of such a book are triangles, squares, circles, spheres, cones, pyramids, and other mathematical figures, most apt for such a reading." In Stillman Drake, *Galileo at Work: His Scientific Biography* (Chicago, IL: University of Chicago Press, 1989).

19 Galileo Galilei, *Dialogue Concerning the Two Chief World Systems, (1632),* trans. Stillman Drake (Berkeley and Los Angeles: University of California Press, 1967).

20 J.-M. Lévy-Leblond, *(Re)mettre la science en culture: de la crise épistémologique à l'exigence* éthique, p.13. Available online at https://www.researchgate.net/profile/Jean-Marc_Levy-Leblond/publications

21 Pauli, ibid., 221.

3. Signs in the Sky:
The Withdrawal of Projections

Projection of unconscious content onto the sky. Search for the ultimate scientific theory and yearning journey of the soul back into Oneness. Re-emergence of the symbol in the interpretation of a theory.

In ancient times, the psychic life of humanity was projected outside itself onto constellations, represented by different mythological figures. Mankind scanned the sky, which held its unconscious content. Humans recognized twelve constellations in the zodiac, this band of space traversed by the planets, also called the 'belt of Ishtar,' the name of an ancient goddess.

Since their Chaldean origins, these constellations – today understood as symbols – reflect an effort to explain, or a dream of bringing celestial geometric order to the earthly chaos. They told myths and legends corresponding to an ancestral substrate of the collective soul. Perceived as beliefs, constellations were considered ridiculous by science, which had demonstrated that they were merely the product of the perspective from Earth.

The characteristic figures of the zodiac have no physical basis; furthermore, the constellation sizes are unequal and – in reality – it is possible to count thirteen of them. With the advent of science, these discoveries have contributed to the arguments regularly trotted out by opponents of astrology. Meanwhile, siderealist astrologers (using the fixed zodiac of constellations) or others, hiding behind the mobile zodiac of seasons, were just as misled by assigning stars a type of "natural scientific influence" on character and destiny.

However, the intuition that myths do not exist in a universe outside the individual, but rather that they come from a spiritual unknown inner world, had gradually arisen, without waiting for the emergence of science. The great Neoplatonic tradition, which had lasted until the Renaissance, described a level of reality whose characteristics were close to universal knowledge shared by all humankind.

Paracelsus – a Swiss physician of the Renaissance – spoke of an inner sky that corresponded to the external sky within a human dual constitution. The outer sky, that of the different zodiacs, referred to the inner sky, which was of greater importance. Several centuries ahead of his time, he considered the existence of the psyche's domain of reality, where the images of the soul reflect the vicissitudes of the human adventure.

Jung was led to the concepts of a collective unconscious and archetypes through his long practice with patients. The famous psychologist used a precise empirical approach to examine and compare phenomena such as dreams, visions, hallucinations, fantasies, etc. He observed the recurrence of isomorphic structures which changed their appearance from dream to dream, but whose functional forms remained essentially the same.

After extensive research on myths, legends, and tales from geographically and historically diverse cultures, he found common thematic structures. He tried several times to clarify the distinction between the archetype and the archetypal images that represent it: "archetypes are not determined as regards their content, but only as regards their *form* and then only to a very limited degree."[22]

The primordial image, or archetype, is only determined when it has become conscious and has, therefore, been filled with the material of conscious experience. Jung's archetypes are to psychology what the *a priori* categories, or Platonic ideas, are to philosophy.[23] To make archetypes understood, he compared them to the axial system of a crystal that, although it has no material existence, displays the crystalline structure in the mother liquid.

Projections are contradictory trends that cannot be understood through reason, and therefore tend to leave the psychic space and embody themselves in external objects. The archetypes, representing unknowable and unrepresentable empty forms, access consciousness in the form of archetypal images – or symbols – specific to personal or cultural experience. In Kantian terms, the archetype is the *noumenon*, while the archetypal image is the *phenomenon*.[24]

Zodiacs most often have a circular or square form. These are archetypal images. We can approximate them with the oriental *mandalas*, whose centered geometric ordering are representations of the order in the universe in its many physical and psychological realms. They express forms arisen from the deepest layers of the human mind and represent the image of the Self which is, according to Jung, the overall ordering archetype of all archetypes.

This instance exists a priori and unifies in one set the entirety of the conscious and the unconscious. In a way, it preforms the 'Ego': "It is not I who create myself, rather I happen to myself," declares Jung.[25] The twelve signs of the zodiac can be considered as twelve archetypal images symbolizing universal psychological experiences, and they refer to unrepresentable archetypes in a movement of circumambulation around the Self.

In their *original phase of invention*, scientific hypotheses appear *also* as projections of unconscious contents, just as are the zodiacs and various symbolic figures of astrology. Theoretical models cannot be formulated otherwise than constrained by the underlying beliefs and worldviews of their inventors. Facing ever elusive reality, nascent proto-ideas are projective elaborations comparable to the symbols of the trinity or to the Platonic polyhedra that Kepler used as a starting point for his new approach to the world. In the same way, cohesive, bonding and belonging aspects of religion are projections of an unconscious power of Oneness that may transcend all forms of spiritual longing for wholeness, of which the quest for the ultimate scientific theory that would unify the four currently known interactions is but another projected form of the same unifying dynamics.

Like all intellectual reasoning, ideas or scientific hypotheses always develop from the soil of imagination. Initially, the physicist conceives of hypothetical images that he specifies in a long process of discrimination and ordering. This process leads to laws expressed in mathematical language as mathematical structures that he hopes will coincide with the manifestations of physical phenomena. At this phase, the mathematical language in itself – the theory or model – no longer has meaning. The image and the symbol that had constituted the charge have disappeared during the formulation phase of the equations. It is only through the interpretation of the model that meaning reappears in the interrogation of emotional commitments and philosophical prejudices.

Albert Einstein, for example, who invented the first modern model of universe in 1917 in the wake of his bold formulation of a space-time curved by matter and energy, could not claim total freedom from metaphysical bias. He believed the universe was *static* – that is, unchanging in time. But he realized that his equations led to an evolving universe. To be consistent with his belief, he then introduced a cosmological constant to counteract the gravitational effect that he was to eliminate later, when observations overcame his prejudices.

The idea of an evolving universe kind of evoked the Genesis narrative and, therefore, had been forcefully repressed not only by Einstein but also by the whole scientific community. It took a long time to catch on and two great scientists deserve full credit for their pioneering work. In both cases indeed, underlying themes related to religious and political prejudices have probably guided the thinking processes and infiltrated what was to become their successful model. Basic convictions of the existence of an arrowed time were deep-rooted in the culture of both scholars within either the Christian idea of providence or in the materialist framework of a theology of progress.

In the year 1927, cosmologist Georges Lemaître was the first to propose a core theory, which would later evolve into the Big Bang model. As he was also a Catholic priest, his religious commitment was probably instrumental in the development phase of the intuitive assumption of the 'primeval atom' which may evoke the idea of creation of the biblical text, but also the idea of expansion from a 'primordial egg' or 'primordial seed' common to many earlier cosmogonies. While methodologically separating scientific approach from theology, he was able to carry out his theoretical work and apply it to the first observations of the recession of galaxies.

Well aware that he was working on a scientific model, he "does not give a definitive status to his cosmogony. For him it is only a hypothesis, open to a possible revision."[26] The link between science and faith was not for him "a question of thinking, but a way to live out scientific action, supported by a motivating hope."[27] Even if he evokes a possible concordance of the initial singularity with the hidden God, the 'Deus absconditus' of Judeo-Christian theology, "the omnipresent divine activity is essentially hidden everywhere. It can never be about reducing the Supreme Being to the level of a scientific hypothesis."[28]

In the late 1940s, the American physicist of Russian origin, George Gamow, transformed Lemaître's 'primeval atom' hypothesis into a sophisticated model of the early universe. He assumed the initial state to consist of a very hot, compressed mixture of nucleons and photons, thereby introducing the hot Big Bang model. Gamow enriched Lemaître's hypothesis by adding the concept of temperature. Expecting a thermal radiation, he predicted the existence of a fossil radiation at a temperature of 5 K which was discovered in 1965 at the temperature of 3 K. During a discussion about scientific representations,[29] Michel Cazenave advances the idea that Gamow was unconsciously stimulated by the culture of historical materialism in which he had bathed in his youth, to carry out the idea of a thermodynamic history of the universe despite a dogmatic opposition favorable to a stationary model.

Ultimately, not only Einstein's, Lemaître's, and Gamow's original models but, as Arthur Koestler asserted, "[...] *all* cosmological systems, from the Pythagoreans to Copernicus, Descartes and Eddington, reflect the unconscious prejudices, the philosophical or even political bias of their authors; and from physics to physiology, no branch of Science, ancient or modern, can boast freedom from metaphysical bias of one kind or another."[30]

However, one should not run too easily to the conclusion that 'scientific truth' has been relativized. Prejudices are not bad in themselves, they just show that physics feeds from the same humus as art. Mathematical models are made by human beings who are immersed in a particular culture, shaped and conditioned by their particular cultural-religious context or background. Once constructed, models are appraised *independently* of their inventors. But when obvious metaphysical claims are served with the model, the antagonism of the scientific community for religion may sometimes reawaken. Such is the case of cosmologist Frank Tipler who started his career as "a convinced atheist,"[31] published technical work on general relativity that were well received by the scientific community, and respected the central concern of a scientific theory, namely the possibility to be falsified. For, although truth is the aim of science, science cannot tell the truth but only establish falsehood. Indeed, Tipler has gradually turned to philosophical biases of such a strength that, after many years of research, he realized that the laws of physics gave him no choice but to be a Christian. In his view, when applied to cosmology, these laws validate the existence of the Judeo-Christian god and theology will become or already is a part of physics! For him, then, the principles of quantum mechanics allow us to calculate, within the realm of sensible reality, the probability of the resurrection of Christ and other infinitely improbable religious miracles.[32] We are left with *one* – and only one – plane of reality, with the laws of physics controlling everything, even inclusive of mathematics, the latter also regarded as a branch of physics. The reception of Tipler's works has suffered from this mixing of genres. His *methodological* approach, however, remains free of theological contamination and permits us to consider his 'Omega Point' as a true scientific model that can be refuted and superseded.[33]

This is an interesting model, simply because of the mere fact that few cosmologists are interested in the future of the universe compared to its past. Moreover, he does not neglect the effect of life and consciousness, even posing the hypothesis that life, once it has appeared, must never be extinguished. As the 'final state' of the universe possesses infinite computational resources, life will be able to resurrect the dead via perfect computer emulation. Finally, his model leads to what he considers to be testable predictions: a closed universe with a very special global causal structure. He writes: "the existence of intelligent life in the far future is required for the consistency of the laws of physics, since in the absence of life acting to speed up fermion annihilation, the universe would accelerate forever, violating unitarity, forcing quantum field theory to diverge, and incidentally, extinguishing life. But if life goes on forever, with the computer capacity increasing without limit, then life in the far future could emulate the present state of the universe down to the quantum state; indeed, in such an emulation, the structure of the world would be a mere manifestation of pure number."[34]

It would not be the first time that a theory of everything describing and unifying all the forces in physics remains out of reach of experimental refutation. But even if his predictions could not be tested, this model would retain its value: that of participating in the coherence of a set of schemes of intelligibility that ultimately could lead to a new model directly confronting cosmological observations.

Like in all human activity, unconscious bias and mythic images enter the creative act that lead to the construction of a cosmological model. Even once these intuitive leaps have been processed by analytical reasoning and produced a mathematical equation, this language made up of abstract symbols does not interest only the intellect. It may arouse aesthetic feelings of harmony in numbers and, when the theory is subject to experiments and *interpreted*, subjective factors, prejudices and bias enter anew into the stage and may trigger controversies. Thus, scientific thinking is irrigated by ineradicable irrational factors with hidden meaning available for interpretation.

"What would, therefore, be given *ante rem* in the idea would be its affective-representative mould, its archetypal motif. This also explains why scientific rationalisms and pragmatic procedure never free themselves entirely from the halo of the imaginary, and why rationalism, or any rational system, carries within it its own fantasms. As Jung says, 'images which serve as the basis for scientific theories keep within the same limits [...] (as those which inspire fairy-tales and legends).'"[35]

Numbers also arouse unconscious resonances. Their ineradicable quality of ordering can re-emerge only through the re-connection that is reconstituted in the interpretation of the theory or model. Their archetypal aspects, which we will later detail as conceptions specific to Jung and Pauli, permeate science.

ENDNOTES

22   Jung, *The Archetypes and the Collective Unconscious*, CW 9, Part 1, trans. R.F.C. Hull (Princeton, NJ: Princeton University Press, 1990), § 155.
23   Jung, *The Psychology of Kundalini Yoga: Notes of the Seminar Given in 1932 by C.G. Jung* (Princeton, NJ: Princeton University Press, 1996), 9. Accessed October 28.2017. https://monoskop.org/images/0/08/Jung_Gustav_Carl_The_Psychology_of_Kundalini_Yoga_1932.pdf
24   Ibid.,10.

25 C. G. Jung, *Psychology and Religion: West and East, CW 11*, 2nd ed., trans. R.F.C. Hull (Princeton, NJ: Princeton University Press, 1969 and London UK: Taylor & Francis Ltd, 1970), § 391.

26 Dominique Lambert, *The Atom of the Universe: The Life and Work of Georges Lemaître* (Kraków, Poland: Copernicus Center Press, 2015).

27 Ibid.

28 Ibid., who quotes Georges Lemaître (1936).

29 In "General discussion following Marc Lachièze-Rey's presentation," *Third Symposium Brussels, 3-6 June 2003, Science and its Representations*, Talk recorded for the programme "L'éloge du savoir" presented by Michel Cazenave and broadcast on France Culture Radio on September 4.2003. See also: Helge Kragh, *Cosmology and Controversy* (Princeton NJ: Princeton University Press, 1999), 262.

30 Arthur Koestler, *The Sleepwalkers: A History of Man's Changing Vision of the Universe* (London: Penguin Classics, 2014), 11.

31 Frank Tipler, *The Physics of Immortality: Modern Cosmology, God and the Resurrection of the Dead* (New York, NY: Doubleday & Co., 1994), ix.

32 Tipler, *The Physics of Christianity* (New York, NY: Doubleday & Co., 2007).

33 The physics of the Omega Point cosmology is defended by atheist physicist – inventor of the quantum computer – David Deutsch. See his book: *The Fabric of Reality* (New York, NY: Viking Adult, 1997).

34 Tipler, *The Structure of the World from Pure Number*, Rep. Prog. Phys. 68 (2005) 897–964 doi:10.1088/0034-4885/68/4/R04.

35 Gilbert Durand, *The Anthropological Structures of the Imaginary*, trans. Margaret Sankey and Judith Hatten (Mount Nebo, Australia: Boombana Publications, 1999), 61.

4. Science Finds its Origins *Outside* of its Own System of Thought: Natural Numbers and the Foundations of Mathematics

Projection of unconscious content onto the sky. Search for the ultimate scientific theory and yearning journey of the soul back into Oneness. Re-emergence of the symbol in the interpretation of a theory. Irrational properties of numbers. Numbers as a precondition for knowledge. Elementary particles know 'how to count.' Unity and the dialectical complementarity of symmetry/ asymmetry.

4.1. Limits in Scientific Knowledge

Science has gradually discovered the structures that make up inert matter and living matter, but making progress requires going back to settle one's foundations. We must rework old concepts to tackle the real with new constructions. In order to systematize the entire edifice of mathematical knowledge, in the early twentieth century, scientists were able to identify some highly abstract first principles, such as the mathematical notion of sets. But, since ancient times, many paradoxes remained in logic, such as the liar paradox.

To overcome these contradictions and paradoxes, a vast formalization plan for the foundations of mathematics was launched. Mathematician Kurt Gödel (Fig. 4) did not believe in the hypothesis of completeness and decidability of sufficiently expressive formal theories. In 1931, while exploring a variant of the liar paradox – which he had converted into a mathematical formula – he proved, to widespread surprise, that the proof of mathematical theorems could not be completely mechanized.

Figure 4. Kurt Gödel (1906 -1978).
(Image reprinted with permission from
IMAGNO Brandstätter Images.)

A mathematical description of the world could not be both consistent and complete: if it were consistent, then it might have omitted some truths, and if what had been omitted were included, then it would become incoherent.[36] Thus, even in the mathematical field, for a long time considered above suspicion, Gödel's theorem had 'logically' demonstrated the limits of logical demonstration. For example, if a mathematical theory is complicated enough to include the natural numbers, then it must contain statements which cannot be proven true or false. They are, in other words, unprovable. For all mathematical theory (and, therefore, for any physical theory), which includes arithmetic, the truths about natural numbers could not simply be deduced in a logical formalism.

Not only Jung and Pauli but also some important mathematicians have recognized about these numbers a certain *irrationality.* Such is the case of Hermann Weyl who remarked in 1949:

"'Mathematizing' may well be a creative activity of man, like music, the products of which not only in form but also in substance are conditioned by the decisions of history and therefore defy complete objective rationalization. (...) it is surprising that a construct created by mind itself, the sequence of integers, the simplest and most diaphanous thing for the constructive mind, assumes a similar aspect of obscurity and deficiency when viewed from the axiomatic angle."[37]

The numbers are "just there." Number has an aspect of just-so-ness, it is an elementary archetype which is "in itself." As Marie-Louise von Franz expresses "[...] the ultimate basis of all mathematics is the series of natural integers (1, 2, 3, etc.) and ... this basis is *irrational.* It can be neither deduced from nor subsumed under any mathematical or logical principles whatsoever."[38]

Outside the restricted circle of mathematicians of his time, Gödel's extremely revolutionary demonstration did not have much impact, as if the general public, unconsciously, already knew the result, or as if Gödel somehow merely rediscovered, using a scientific method, what everyone already sensed; namely, the feeling of the existence of *several* levels of reality. This ran contrary to what nineteenth century science had asserted in advocating the existence of *only one* level of reality; that of the empirical reality it was studying.

Besides, this is what Gödel's era regarding science (i.e., quantum physics) was in the midst of re-discovering in the core field of scientific activity. After progressing through leaps and bounds toward a totally unified description of reality in a zero-dimensional model, scientists of the epoch were facing the paradoxes of quantum theory that challenged conventional classical thinking. The world of atoms could not be represented by models familiar to the senses and had become quite alien to the existing philosophical approaches. "It was as if the ground had been pulled out from under one, with no firm foundation to be seen anywhere, upon which one could have built."[39]

Thus, Niels Bohr and many other inventors of quantum theory decided to explore other ways of thinking that would help to understand the paradoxical phenomena of microphysics. Discovering limits in scientific knowledge led to the fundamental question: what is reality? With the existence of undecidable propositions, Gödel was doing nothing more than rediscovering through mathematics that scientific knowledge originates *outside* its own system of thought in an 'unknowable,' *other* level of reality.

But the steamroller of scientistic ideas that had dominated throughout the nineteenth century is still present today, as evidenced (for example) by the reductionist project of Frank Tipler to bring all disciplines into physics, confusing a religious symbol with a scientific representation in the illusion of a *single* level of reality. For its construction, science had fed on ancient philosophical or religious traditions, throwing away the ladder once climbed. This was fortunate, because

> "scientific rationalism, whose influence continued to grow from the seventeenth century on, had (*regarding projections of the psyche onto material reality*) a huge merit: it continued to enlarge the gap between Father/Mother opposite archetypes, from intra-psychic mind and outside matter." Science, in this view, has gotten rid of projective animism, and constantly tells us, in an increasingly precise way, what is forbidden to claim regarding the reality of *Being*. [...] All rational effort, all scientific achievement *compels* us to abandon the delights of magic, and instead, forces us to try to understand them, not in

their immediate assertions that have become in law inadmissible but in *what, on another level, and in another context – in psychological context – these assertions try to say.*[40]

But the gap, by dint of being widened, became an abyss. The collapse of the mediatory notion of soul and the clear separation between subject and object led to a surviving claim that the only existing level of reality consists of material substance. This is a refusal of plurality and differences. The world is one, but is no longer *unus mundus*. A 'symbol' has become synonymous with a purely conventional and unambiguous thing that does not belong to the original *sumbolon*, meaning 'to bring together.' Indifferently used as sign, signal, mathematical variable, or image of a dream, it shows a compression in thinking. Once a symbol becomes a mathematical concept, it no longer holds meaning, unlike a true symbol that refers to something other than itself in *another* plane of reality. It would be reasonable to grant an epistemological dignity to traditions that allowed the adventure of science – like alchemy, which had fallen into oblivion – or others, qualified as para-sciences – like astrology or numerology, which had been abandoned to charlatans.

4.2. Levels of Reality and Mirroring Effects

These traditional subjects may sometimes tally with certain descriptions or structures of thought in modern science. Analogies have long been woven between what is above (the macrocosm) and what is below (the microcosm). To the extent that these analogies involve scientific knowledge, it is important to be aware of the difference between levels of reality. Otherwise, even if poetic correspondences are to be found, they would not contribute to the reciprocal comprehension of the disciplines placed into relation.

The fundamental distinction is that between *Being* ("l'Être," reality "in-itself") and *beings* ("l'Étant," the shimmering phenomena). The aim of science is not knowledge of Being, but that of beings. Delusions and fantasies arise from the confusion of these two levels of reality. Indeed, within science, quantum physics has brought to light objects that behave completely differently from their classical counterparts, such as in the particle-wave duality or in the entanglement of particle wavefunctions.

In his book *Science and the World Soul*, Michel Cazenave tries to reassess a fundamental unity of the world and of mankind by considering the notion of Being in an essential dialectic between unconscious knowledge and rational processes. Based on the four-pointed logic of the tetralemma that expands the dilemma (logic of the excluded middle), he distinguishes four levels of reality (summarized by me in Table I). This enables him to posit the idea of a 'differential unity,' a principle of coherence in which all disciplines are respected regarding their objects of study and their own methodologies.[41]

Logical alternatives	Levels of reality	Table I
(the excluded middle) A is A and A is not B	Beings or entities	Level of the psychology of the conscious and of phenomena that are relatively separate, and studied by classical macroscopic physics.
	The Totality-of-beings	Level of the total psyche personified in an individual, of phenomenal relativist physics and phenomenal quantum physics, of David Bohm's 'implicate order'. Particular entities ('explicate order') originate from these actual globalities.
(the included middle) A is A and B	The Plane of Being	Realm of the potential totality of the objective transcendental psyche, of the *unus mundus*, of David Bohm's 'super-implicate order,' of the real at the roots of quantum physics where paradoxes and complementary pairs are found (conscious-unconscious, wave-particle).
A is neither A nor B	Being	Unknowable and imparticipable, beyond all contradiction and all identity.

The excluded middle, inherent to Aristotelian logic, has been employed for over two thousand years. It perfectly accounts for the first two levels, namely 'beings' and 'totality-of-beings.'

The tetralemma, or 'catuskoti' in Sanskrit, admits a first additional possibility (the included middle). It corresponds with the 'Plane of Being', where the soul is grounded as intermediary, which takes into account the co-existence of contradictory states such as those that have emerged in depth-psychology or quantum physics. The fourth logical position points to Being itself, i.e., "negation and negation of negation, of which nothing can be said that is not ascent, alienation, or nostalgia."[42]

Yet, such a hierarchical structure cannot go without links – or else how could one access the thought of Being, participate in the Imparticipable, in the transcendent Absolute without the latter forfeiting its radical transcendence? Cazenave proposes the symbol of the mirror as far as

> [...] one is willing to admit that each level is existentiative of the level that follows it, and that each 'higher' level necessarily participates de facto in its lower level by founding it in the same time that it exceeds it, or that, as Shayegan puts it, "the higher plane is *reflected* in the lower plane which is like its actuated epiphany [...]. Each lower level is the actualization of the virtual realities of the higher plane in the same way as it potentially contains the act of the level that ontologically follows." As in a multifaceted mirroring system, in some way "each mirror has a face turned towards that which is superior and that it reflects, just as it illuminates the degree which is immediately inferior." From level to level, Being is thereby reflected, and at the same time, the Being dwells up to the realm of sensory existence – although it absolutely transcends it in its successive foundations. In other words, *all scientific subjects say something of the Being at their own levels.*[43]

4.3. Qualitative Number and Quantitative Number

Long before Gödel and the collaborative works of Jung and Pauli, many ethnological studies, as well as philosophical or historical studies of religions, had already shown that numbers expressed not only functions of enumeration and calculation, but also metaphysical ideas indicating a qualitative cosmic order. For the ancient Greeks, the art of calculation must:

> [...] not be treated in the manner of the populace, but in a way which directs men to contemplation of the *essence* of numbers, not from the viewpoint of commerce as with merchants and peddlers, but for the good of the soul in facilitating for it the means of elevating the order of transitory things towards the truth of being.[44]

One of the fundamental features of Chinese thought, Marcel Granet writes, is "an extreme respect for the numerical symbols that combine with extreme indifference to any quantitative conception."[45] The ancient Chinese perceived numbers rather as symbols governing a variety of areas of cultural and intellectual life. They were indicators of the quality of temporal phases and unveiled intentions of Tao; namely, development opportunities relevant to a given situation.

This view is illustrated by Granet through the following story:

> The *Zuozhuan* records the debate of a council of war: should the enemy be attacked? The leader is tempted by the idea of combat, but it is necessary for him first to consult his subordinates and take account of their advice. There are twelve generals at the council. The advice is divided. Three leaders refuse to engage in combat; eight of them want to go to battle. The latter are the majority and proclaim as much. The advice that unites eight voices however does not override the advice that unites three of them. Three is almost *unanimity*, which is a very

different thing than the *majority*. The general-in-chief will not fight. He changes his opinion. The advice to which he adheres by giving his *single* voice is then imposed as a *unanimous* opinion.[46]

This little story expresses very well the role of number in an ancient society. Here, inner and outer worlds were seen as a cyclical flow of energy, and numbers appeared naturally connected with the idea of meaningful coincidences (synchronicity), reflecting the union of psychic being with matter. This thought is hardly understandable to western civilization, marked by the prejudice according to which a number can only designate a quantity.

The correspondence between Jung and Pauli demonstrates their close cooperation in the effort to approach the definition of archetype in general, and number in particular. Neutral with regard to psycho-physical distinction, number is for Jung "[...] the most primitive element of order in the human mind, seeing that the numbers 1 to 4 occur with the greatest frequency and have the widest incidence. In other words, primitive patterns of order are mostly triads or tetrads."[47] From the psychological point of view, he defines number "as an archetype of order which has become conscious."[48]

Pauli said that number was a "primary probability," thereby referring to the link he perceived between the archetype and the concept of probability, which appears in science with the transition from an unknowable probability in quantum reality to a discrete physical form in classical reality. In this passage, the wave function, which represents all the possibilities of the superposition of quantum states, 'collapses' in an irreversible operation. It is only by calculating probabilities that quantum physics remains deterministic and conserves the law of causality.

Jung tried to use statistics to prove his concept of 'synchronicity.' This concept brings together phenomena of meaningful coincidences (the meeting between a qualitative psychic event and an objective material fact), including correlations involved in traditions such as the *Book of Changes* (the *I Ching*), the Tarot of Marseilles, or astrology. Yet, Pauli did not support Jung's synchronicity, and refused to place these correlations on the same level as those involved in physics, because the former involved subjectivity, especially regarding the meaning given to the event by whomever it is that notes the coincidence.

In a letter to Jung on 12 December 1950, Pauli writes:

> In cases of non-psychic acausality, on the other hand, the statistical result as such is reproducible, which is why one can speak here of a 'law of probability' instead of an 'ordering factor' (archetype). Just as the mantic methods point to the archetypal element in the concept of number, the archetypal element in quantum physics is to be found in the (mathematical) concept of probability – i.e., in the actual correspondence between the expected result, worked out with the aid of this concept, and the empirically measured frequencies.[49]

Pauli ties archetype and number to the notion of probability when he writes: "In physics however we do not speak of self-reproducing 'archetypes', but of 'statistical laws of nature involving primary probabilities;' but both formulations meet in their tendency to extend the old narrower idea of causality (determinism) to a more general form of 'connections' in nature, a conclusion to which the psycho-physical problem also points."[50]

4.4. The Number-Archetype, the Point of Contact between Matter and the Psyche

Jung and Pauli understood that they could not define the archetype. They had been led to posit it as originating from a hypothetical potential instance. It was to be tackled only from a multilayered approach to reality, consistent with the spirit of Gödel's demonstration. Being invisible and outside of any language, they could only approach it while inevitably missing it at every attempt. That explains Jung's continual variations in his definitions of the archetype: from his initial conception of "primordial images" to the idea of *arche* (the source) and its dual aspect of foundation and principle.

The archetype, as the collective unconscious, can only be understood as *representations*: "When I say 'atom' I am talking of the model made of it; when I say 'archetype' I am talking of ideas corresponding to it, but never of the thing-in-itself, which in both cases is a transcendental mystery."[51]

Very early, Jung had the intuition of the 'psychoid' nature of the archetype, due to its phylogenetic and biological foundations. He considered several strata in the collective unconscious: first the family collective unconscious, then the collective unconscious of the ethnic and cultural group, and finally the primordial collective unconscious. In the latter is found all that is common to humanity, such as the fear of darkness and the human survival instinct.

The archetype is thus what organizes and structures all the mental processes of the human being, as well as: "[...] the impulse of birds to build nests, or ants to form organized colonies."[52] The archetypes would thus be analogous to instincts, which – in ethology – are predispositions to action. In Lorenz's geese experiment, the lure that triggers the relation response showed that the young goose "[...] has been endowed with an *a priori structure* filled by experience: an empty form, in fact, which accurately *informs* its way of being in the world."[53] In animals, these structures are closed, while in humans – largely deprived of survival instincts – archetypes are not rigid.

The immaturity of human beings (neoteny) allows for reflective mediatization of the brain, which thus becomes cultivable over the long years of childhood. Many biologists and zoologists

> have shown that the Jungian theory of archetypes might provide a suitable foundation for an overall view of human and animal psychology. Lorenz speaks of 'innate schemata' (that is, certain forms of "inborn reaction to characteristic stimulus situations"); these schemata are "independent of experience", and in them a "formal similarity to certain human relationships based on inborn schemata may be observed also in animal behavior." He stresses that by this he does not mean 'innate images' but, rather, the 'preformed potentiality' of such an image, and declares that it is "experience which fills the form with matter," and also that "certain types of human reaction cannot be explained by specific adaptation or expediency for the preservation of the race, but are direct manifestations of laws which attach to all living creatures as such … and which seem to be given *a priori*." [54]

Thus, archetypes represent the openness of instincts. *A priori* structures allowing for representation, they are rooted in somatic processes of the nervous system of the body and form the 'bridge' with matter. Pauli, meanwhile, in his research on Kepler, will find that the latter attributed to Proclus – his favorite author – the idea that formal concepts are derived "[…] from a natural instinct (*instinctus*) and are inborn in those beings as the number (an intellectual thing) of petals in a flower or the number of seed cells in a fruit is innate in the forms of the plants." [55]

Thus, numbers refer to a background of reality where psyche and matter are undifferentiated. Today, it is well established that infants can tell the difference between small numbers such as 3 and 4, without learning. This faculty was also detected in crows and even in primitive creatures such as fish or insects.

Why stop at a certain level of evolution? If an archetype is primarily an unconscious disposition, its engram is not only inscribed at the biological level, but also within the so-called inert matter. Marie-Louise von Franz (Fig. 5), a close collaborator and follower of Jung, imagines that:

> [...] the most elementary particles, such as quarks, protons, mesons and baryons, 'know how to count.' They combine in hexagons, triplets, octuplets, and so forth. Particles would not know how to count as we do, but would be more like a primitive shepherd, who, without knowing how to count beyond three, can tell in the twinkling of an eye whether his herd of 137 animals is complete or not.[56]

And she adds that "man possesses an unconscious 'numerical sense,' and this is probably the sense that subatomic particles possess. Starting with numbers in the subatomic dimension, it would be a long road before reaching the first unicellular living beings with its genetic programing, which is also numerical."[57] Archetype acts as a bridge with matter and, insofar as it is an archetype, number is

Figure 5. Marie-Louise von Franz (1915-1998).

thus not only a psychological factor but, more generally, a factor which orders the world.

4.5. Symmetry and Asymmetry in Number

During the mathematical construction process, number has lost much of its rich aura of mystery. Alchemy considered odd numbers as masculine and even numbers as feminine. However, among these qualitative aspects related to the symbol of the union of sexes and the engendering of the One – possessed by number before its reduction to mathematical number – parity is all that remains today.

Symmetry has always played a fundamental role in physics, especially since the early twentieth century when Emmy Noether demonstrated that any property of symmetry corresponds to the conservation of a certain physical quantity: namely, an invariant allowing for the development of new laws. Scientists have therefore been interested in symmetries present in spaces more abstract than ordinary space.

Like almost all physicists, Pauli thought that all the laws of nature would obey the three CPT symmetries. According to von Franz, he considered symmetry "as a form of God" and – with the approaching results of an experiment expected to show that weak force did not respect parity symmetry – he was even willing to bet a large sum, asserting he did not believe that God would be "a weak left-hander."

Like all his other colleagues, on 15 January 1957, Pauli was stunned when parity – which is to say, the left-right exchange or mirror reflection – was announced to be overthrown for the fourth known force of physics. The August 1957 correspondence with Jung that followed this event is very interesting because it shows that:

> 1) the problem of symmetry and asymmetry coincided with similar concerns for Jung, who, during the same period, had written his essay on UFOs, interpreting these as projections of the individuation process. Jung believed that unconscious manifestations, of which flying saucers are an example, "[...] point to a 'slight left-handedness in God,' in other words, to

a statistical predominance of the *left* – i.e., to a *prevalence of the unconscious*, expressed through 'God eyes,' 'creatures of superior intelligence,' intentions of deliverance or redemption on the part of 'higher worlds' and the like."[58] Jung saw the conscious-unconscious asymmetry as a third element, "an *archetype* that could unite or reconcile the opposites."[59]

2) In 1954-55, Pauli had been interested in mathematical problems of reflections. He recalled "a very impressive dream that occurred after I had finished my work, work which had struck me as a thoroughly straightforward activity." In his dream returned "the 'dark woman,' namely the *Chinese woman* (or the Exotic One) with the typical slanted eyes" who had already appeared in previous dreams (letter of 27 February 1953) and moving in a rhythmic dance, "the index finger of her left hand and her left arm pointing upward, her right arm and the index finger of her right hand pointing downward."

Pauli had recognized her as a female figure (the anima archetype) in relation to the pairs of opposites. While dancing, she had invited him to stand on a platform to talk to strangers but, in doing so, he woke up. The dream of 27 November 1954 is even more impressive, because the 'Chinese' or 'dark woman' is with him in a room where experiments are being carried out in which 'reflections' appear. Other people are in the room. They regard these reflections as 'real objects,' whereas, he and the 'Chinese' know that they are just 'mirror images,' but this shared secret 'fills us with *apprehension*.' Afterwards, they walk alone down a steep mountainside.

While, in keeping with the eternal Ideas, Pauli and his fellow physicists had remained prisoners of the Platonist heritage of symmetry, that dream dating back from three years before showed that Pauli had unconsciously foreseen asymmetry in physics. But he had remained in the mist of the unconscious, despite, perhaps, a knowing wink through the ages from his predecessor Kepler.

"I felt as if I had been awakened from a sleep ..."[60] Kepler, his illustrious predecessor, did write this, three centuries earlier, after six years of work when, finally holding the secret of the orbit of Mars, he had resigned himself to a 'lesser symmetry.' Instead of God slightly left, he had faced an 'out-of-shaft spinning God,' which is to say: the loss of the ideal of symmetry in calculating an orbit. He had to break one of the Aristotelian dogma which claimed to represent all the movements in the sky with perfect figures: spheres and circles. He understood that the movement of Mars was not a circle but, to his horror, an ellipse that he saw "as a cartload of dung."[61] Yet the recognition of a lower symmetry in the 'degenerate circles' will become his first law that will prevent complications of previous systems.

Along the same lines, Proclus, born in 412 A.D., not only rejected Ptolemy's epicycles and deferents that were to be solved by Kepler's ellipse in a heliocentric world, but he also rejected Ptolemy's interpretation of the precession of the equinoxes, today perfectly understood by Newtonian mechanics. This phenomenon originates in the wobble motion of the orientation of Earth's axis of rotation, in the opposite direction of the spin with a period of 26,000 years. Consequently, every year the position of the Spring equinox is slightly shifted. This motion contradicted Plato's dialogues and the ancient science of the Chaldeans who, in Proclus' view, would certainly have discovered this phenomenon if it were real and replaced the sidereal (fixed) zodiac of the constellations with the mobile (tropical) zodiac of the signs. Thus, Proclus, could not be satisfied with such additional complications as a shifting of all the stars with respect to the celestial sphere. These 'ad hoc hypotheses' (not necessary false) could not reach the essence of things. Like Kepler and Pauli, he lived in an *ideally created symmetrical world.*

Getting back to Pauli, far more than just a presentiment of asymmetry, his dream of 27 November 1954 seems to have had a premonitory aspect today. In fact, Dr. Chien Shiung Wu, the physicist who made the experiment leading to the proof of the violation of parity, was a beautiful Chinese-American woman.[62] And, this 'Chinese influence' became a real 'Chinese revolution' because the two theoretical physicists who had suggested the non-conservation of parity for weak interaction – after many attempts, ignored by their colleagues – were two Chinese-American young physicists: Tsung Dao Lee and Chen Ning Yang.

According to Martin Gardner: "[...] perhaps the familiar asymmetry of the Oriental symbol, so much a part of Chinese culture, may have played a subtle, unconscious role in making it a bit easier for Lee and Yang to go against the grain of scientific orthodoxy, to propose a test which their more symmetric-minded Western colleagues had thought scarcely worth the effort."[63]

Neils Bohr, the quantum mechanics pioneer behind the 'complementarity principle', had chosen this Yin-Yang symbol as a crest. Illustrating the complementary relationship between the wave and the particle, its deeper emerging harmony, mixed with subtle asymmetry, seems to reflect the eternal change of nature that is increasingly difficult to grasp due to a mode of thinking that had remained attached to the ideal Greek spirit of symmetry.

Like many fellow physicists, Pauli was very affected by the news and, according to Prof. C.A. Meier, "[...] happiness was destroyed when the concept of the conservation of parity was shattered. He felt betrayed in his love of symmetry, but his peace of mind was gradually restored as he buried himself in his work."[64] It seemed increasingly clear that in a completely symmetrical world, homogeneity and monotonous uniformity would have reigned, preventing the appearance of objects or phenomena. Yet, already in 1848, Louis Pasteur's work on 'enantiomers' had showed that asymmetry was a given characteristic of living matter.

Since the Jung and Pauli exchanges on asymmetry, transdisciplinary qualitative notions like "spontaneous broken symmetry" have appeared in science, often associated with the notions of attractors. Broken symmetry means disorder and contingency, while symmetry implies order and regularity. The world exists only through the unity and dialectical complementarity of symmetry/asymmetry, beautifully illustrated by the complex union of the Yin-Yang symbol.

From the point of view of mathematical symmetry, the Yin-Yang symbol is part of the 'rosettes group.' It is not superimposable onto its mirror image, but it has a dual symmetry of rotation around the center point, combined with color inversion. The dividing line, formed by a curve in two semicircles with the small black circle in the white part and the small white circle in the black part, evokes the impossibility of the pure binary, while the subtle asymmetry of the mirror evokes an eternal flow (Fig. 6).

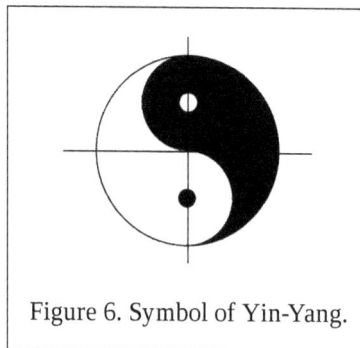

Figure 6. Symbol of Yin-Yang.

A quick analogy could associate the small black and white circles to attractors, while the unstable equilibrium points (as much white as black) on the horizontal line, may reflect the breaking of symmetry. In Taoist cosmology, the asymmetry in the eternal flow of the one and the multiple appears through a dialectic of the first integers, expressed in this excerpt from the *Tao-te Ching* of Lao-tzu:

"The Tao begot one.
One begot two.
Two begot three.
And three begot the ten thousand things.
The ten thousand things carry yin and embrace yang.
They achieve harmony by combining these forces."[65]

This subtle asymmetry, present at the heart of the eternal flow of Tao, is involved in issues related to psycho-physical asymmetries that Pauli and Jung were preoccupied with during their August 1957 exchanges. Jung remarked: "The fact that it is precisely the *weak* interactions that exhibit asymmetry forms an almost comic parallel to the fact that it is precisely the infinitesimal, psychological factors, overlooked by all, that shake the foundations of our world."[66]

In fact, were the laws of physics completely symmetric, we might not be here. A universe created with an equal number of particles and antiparticles would have simply annihilated itself at its very beginning. But a slight flaw in the properties of matter and antimatter, including Charge-Parity symmetry violation, may account for the preponderance of matter over antimatter, and thus for our world made of matter and not just radiation alone.

ENDNOTES

36 John W. Dawson, Jr., *Logical Dilemmas. The life and work of Kurt Gödel* (Wellesley, MA: A.K. Peters, Ltd. 1997).

37 Hermann Weyl, *Philosophy of Mathematics and Natural Science*, trans. Olaf Helmer (Princeton, NJ: Princeton University Press, 2009), 219.

38 M.-L. von Franz, "Some Reflections on Synchronicity," in *Psyche and Matter*, trans. Michael H. Kohn (Boston MA & London: Shambhala, 1992), 255.

39 Manjit Kumar, *Quantum: Einstein, Bohr, and the Great Debate about the Nature of Reality* (New York, NY: W. W. Norton, 2009).

40 Michel Cazenave, *La science et l'âme du monde [Science and the World Soul]* (Paris: Albin Michel,1996), 113, who quotes von Franz, *Psyche and Matter*, 157. [translated by the author]

41 Ibid., 69-72.

42 Ibid.

43   Ibid., 71, who quotes Daryush Shayegan, *Qu'est-ce qu'une révolution religieuse?* (Paris: Albin Michel, 1991).

44   Theon of Smyrna, *Mathematics Useful for Understanding Plato*, trans. Lawlor (San Diego, CA: Wizards Bookshelf,1979), 3.

45   Marcel Granet, *La Pensée chinoise* [Chinese Thought] (Paris: Albin Michel, 1999), [note n° 230 translated by the author.]

46   Ibid., 248. *Round and Square Tuesday, April 17, 2012,* accessed October 28.2017, http://robert-lafleur.blogspot.fr/2012/04/la-pensee-cyclique-9chinese-thought-and.html.

47   Jung, *Synchronicity: An Acausal Connecting Principle,* § 870.

48   Ibid.

49   Pauli to Jung December 12, 1950, in C.A. Meier, ed., *Atom and Archetype: The Pauli/ Jung Letters, 1932-1958* (Princeton NJ: Princeton University Press, 2001), no. 47P, p. 64.

50   Pauli, "Ideas of the Unconscious," in *Writings*, 164.

51   Jung to H. Haberlandt, April 23, 1952 in *Letters 1951-1961,Vol. 2*, ed. G. Adler with A. Jaffé (Princeton NJ : Princeton University Press, 1975), 54.

52   C. G. Jung, "Approaching the unconscious," in Jung and von Franz, et al., *Man and his symbols* (Garden City, NY: Doubleday,1964), 67.

53   Michel Cazenave, "Synchronicité, physique et biologie [Synchronicity, Physics and Biology]," in Cazenave et al., *La synchronicité, l'âme et la science* (Paris: Albin Michel, 1995), 37. [translated by the author]

54   Jolande Jacobi, *Complex, Archetype and Symbol in the Psychology of C.G. Jung*, trans. R. Mannheim (Princeton, NJ: Princeton University Press, 1959), 42. She quotes Konrad Lorenz, "Die angeborenen Formen möglicher Erfahrung [The innate forms of potential experience]," in *Zeitschrift für Tierpsychologie 5*, 1943.

55   Pauli, "The influence of archetypal ideas," in *Writings*, 227.

56   von Franz, *Psyche and Matter*, 256.

57   Ibid.

58   Jung to Pauli, August 1957, in C.A. Meier, ed., *Atom and Archetype*, no. 77J, p. 167.

59   Ibid.

60   Koestler, *The Sleepwalkers*, 337.

61   Ibid., 397.

62   Pauli had professionally known Chien-Chiung Wu for a decade and, according to his assistant Charles Enz, was very much attracted to her. See Charles P. Enz, *No Time to be Brief: A Scientific Biography of Wolfgang Pauli.* (New York, NY: Oxford University Press, 2002).

63   Martin Gardner, *The New Ambidextrous Universe: Symmetry and Asymmetry from Mirror Reflections to Superstrings* (Mineola, NY: Dover Publications, 2005), 221.

64   Meier, ed., *The Pauli/Jung Letters*, foreword, lviii.

65   Lao Tsu, *Lao Tzu: Tao Te Ching*, trans. Gia-Fu Feng and Jane English (New York, NY: Vintage Books, 1989), 44.

66   Jung to Pauli, August 1957, ibid., 168.

5. The Concept of Number-Archetype and the First Four Integers

Numbers as activated points of a field and fundamental dynamic ordering patterns. Three as the coincidence of opposites. Four, ordering scheme par excellence, and archetypal foundation of the human psyche.

5.1. One Becomes Two, Two becomes Three ... Number as a Field

After long observations of facts, empirical data, and reflections on the meaning of science, psychiatrist Jung and physicist Pauli came together in proposing the hypothesis of a potential unity of matter and psyche. This unity transcends the material and psychic realms and speaks the 'neutral language' of number. Modern physics is engaged in a process of unification of which the currently prevailing cosmological 'Big Bang' model is the provisional result. It is interpreted as a narrative which has aggregated geological history and the evolution of living beings. It is a scientific model of wholeness, developed from preconscious phenomena of psychological projections onto the unknowns of matter and cosmos. Made of numbers and equations, it uses a mathematical language, which ultimately rests on irrational data rooted in archetypal number.

In her book *Number and Time: Reflections Leading Toward a Unification of Depth Psychology and Physics*, Marie-Louise von Franz presents numbers not as divisions of the monad but rather as activated points of a field: Whereas numbers above the threshold of consciousness appear to be quantitative discontinuities and qualitative individual numbers, in the unconscious they interpermeate and overlap (as do all the other archetypes of the collective unconscious), participating in the one-continuum which runs through them all. From this viewpoint, all numbers are simply qualitatively differentiated manifestations of the primal one. The latter is a mathematical symbol of the *unus mundus*, isomorphic to the collective unconscious.[67] (Fig. 7)

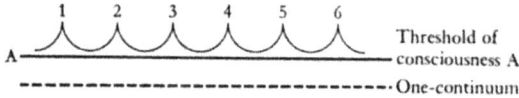

Figure 7. The mutual contamination of numbers.
(Image reprinted with permission from Northwestern University Press.)

In this perspective, two would not be a halved monad, but instead the *quality of symmetry* of the 'one-continuum' equivalent to Tao or the alchemists' *unus mundus*, always equal to itself and beyond all duality. Three would thus arise as the symmetrical axis in the bipolarity of the primal one. This step from two to three is a retrograde process, a reflection leading from two back to the primal one, in which the duel aspect, however, remains contained. And so on, this process can in principle be repeated about all the other numbers. The transition from three to four, made famous by the alchemical formula of Mary the prophetess, will be treated in detail later.

Jung was particularly interested in the number four. He had long reflected on the archetypal quaternity in alchemy, Far East philosophies, Christian theosophies of the Middle Ages, and in the dreams of patients such as the young physicist: Wolfgang Pauli. His 'world clock' dream – a three-dimensional cosmogonical mandala or a modern version of the great vision of Ezekiel – often refers to the number four: four cardinal points personified by four little men, or in other of his many dreams, with four children or four seasons.

As activated points of a field and like all other unconscious contents, numbers are contaminated, i.e., influenced by all the others. They all link up with each other and are interwoven. The number eight, for example, pertains, as the number four, to the wholeness of mandala. The number nine (word very close to 'new' in many languages) evokes a 'new' start. It is thus possible to keep within the first four integers which, ordered by the One-Self archetype, "weave a network of cosmic and spiritual analogies and themes, a circulation in all directions of symbols that all refer to each other."[68]

5.2. The First Four *Integer* Numbers as Fundamental Dynamic Models

For Pythagoreans, the first four integers had a key role, above all the number four. In fact, from the four stretched strings of the earliest version of the lyre (tetrachord) three intervals and four notes were derived, which are the foundation of all musical theory of the ancient Greeks. The first stretched string is 1; the second has a length which is 3/4 of the first, the third 2/3 and the last one, half. When we successively clamp the strings we hear Do, then the fourth of Do = Fa, then the fifth of Do = So, and finally the Do at octave.

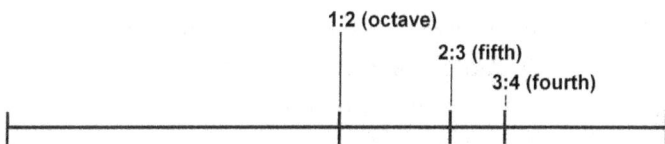

Figure 8. Division of a string.

Tetraktys (four) symbolized the sacred decade (1 + 2 + 3 + 4 = 10) and was represented by four overlapping lines: the first represented zero-dimension (one point), the second, one dimension (a line defined by two points), the third, two dimensions (a plane defined by a three-pointed triangle) and the fourth, three dimensions (a tetrahedron defined by four points). (Fig. 9)

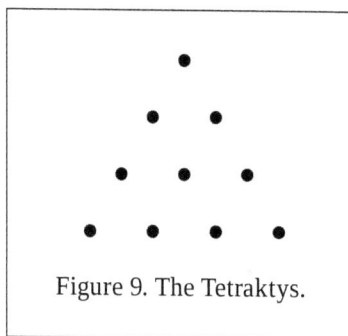

Figure 9. The Tetraktys.

In Pythagorean tradition, integers are developed by successive subdivisions of the One, or Monad, which is considered the first principle from which all other numbers ensue. One is associated with the point, and Two, or the Dyad, is associated with the line. Point and line are two geometric figures without depth and so not concrete, thus explaining why One and Two have been considered as principles rather than numbers.

Three (Triad) and Four (Tetrad) seem to detach from the set of integers with the Greeks as well as with Jung and Pauli. Three symbolizes a dynamic totality: beginning, middle, and end of a process; and Four is the figure of a static wholeness.

Marie-Louise von Franz has described numbers not merely as static forms, but also as vibrational energies. The Pythagoreans had already recognized these energies in the intimate link between numbers and musical tones. She writes:

> Since today we see processes everywhere rather than structures or static orders, I have also proposed seeing

numbers in this perspective – *as rhythmic configurations of psychic energy* [...] and – as I would add today – of psychophysical energy. As a numerical equivalent of the *unus mundus*, I proposed the term *one-continuum*, in which all numbers (including unity) would be configurations of rhythm.[69]

The One or Monad

The One dwells in an ineffable mystery; it expresses the idea of unity. This is the state of the undifferentiated, the completeness, or archetypal wholeness before creation and the subsequent separation of all things. It also symbolizes the distinction of the environment, the beginning, the starting point of all creation when individuality must differentiate into parts. It engenders multiplicity by successive divisions of unity.

Two or the Dyad

Two involves separation, division, or dissociation from the original unity through discord and conflict. The result is a tension of opposites between the One and the Other. With Two, the ability to distinguish one thing from another is born, as well as the possibility of perceiving contrasting oppositions as thesis and antithesis, good and evil, darkness and light, passive and active principles, spirit and matter, black and white, the limited and the unlimited, the one and the many, male and female, stillness and movement, etc. The level of Two is that of the principle of non-contradiction: something cannot be itself and its opposite at the same time.

This is also the principle of all knowledge, the *Nous*: without the observer and their reflective consciousness, there would be no science. More specifically, two is associated with the *threshold* of consciousness that is reached by psychic content which, through the tension of opposites, drives differentiation and conscious realization. In physics, a new interpretation of the 'wave function collapse' mirrors this threshold-like onset of a reduction of a global, infinite, but diffuse knowledge into a determined and precise field of perception. This type of collapse or reduction, which is supposed to provide meaningful conscious moments, is a theory proposed by physicist Roger Penrose who believes that such moments are influenced by 'Platonic values' embedded in the structure of reality.

Three or the Triad

Von Franz sees Three as: "[...] the interference of an observing consciousness which inserts a symmetrical axis into the two-rhythm or else 'counts' the latter's temporal and spatial succession. In terms of content the number three therefore serves as the symbol of a dynamic process."[70]

The masculine principle at the root of action, Three is the second birth of One from Two. It is well reflected in the Heraclitean enantiodromia which describes the effect of a reversal of a unilateral reality into its opposite when it is pushed to its limit.

In Taoism, where things are dynamically connected through the opposition of Yin and Yang, the enantiodromia corresponds to the process by which Yin is transformed into Yang, when it has been pushed to its limit. Similarly, Yang re-transforms back into Yin when it has become an old Yang. The tension between the two opposites is a source of energy, all the greater as greater tension is applied.

Science deals with the enantiodromia phenomenon in different fields of knowledge. It takes the form of moderation laws, meaning that effects are opposed to causes, as in Newton's Third Law of Motion (to every action, there is always an equal and opposite reaction.) In the context of Newtonian dynamics, a force is defined as a change in momentum and therefore, Newton's Third Law (1729) is strictly equivalent to the conservation of total linear momentum of a system before and after a collision. This means that the action-reaction forces appear *simultaneously* in pairs and that *no delay* elapses between action and reaction. But gravity does not act instantaneously as Newton thought. The gravitational waves discovered in 2016 propagate at the finite speed of light. Thus, in the context of the universe at large, there is a *delay* between action and reaction.

In electromagnetism, Faraday's law of induction (1831) describes the macroscopic phenomena of electromagnetic induction where – by virtue of its effects – the induced current is opposed to the cause that gave rise to it. For example, the increasing flux Φ (t) – which is the product of the magnetic field B by the surface – induces an electromotive force e with a polarity such that the induced current generates a magnetic field that will oppose the initial magnetic field B, thereby leading to attenuation of flux Φ (t). Heinrich Lenz (1804–1865) is also credited for having clearly stated the direction of the induced current. Here again, action-reaction is not instantaneous. Time delay arises from the finiteness of the speed of electromagnetic waves (light). If the size of the object is smaller than the wavelength, propagation effects are negligible, and delay is due to its conductivity and magnetic permeability.

Similarly, in a limited chemical process where limitative reactants have not completely disappeared at the end of the reaction, the Law of Mass Action or 'Le Chatelier's principle' (1884) governs the reaction's equilibrium. Any addition of more reactant substances, or decrease in product substances, will shift equilibrium in a forward direction. Conversely, any decrease in reactant substances, or increase in product substances, will shift equilibrium in the opposite direction. Even more simply: around equilibria, a shift of the concentration of a reactant in one direction causes a production or a consumption in the other direction by other reactants.

The Three. Conjunction of Opposites and Dialectical Process.

The theme of the struggle of opposites runs through the history of ideas, taking different forms according to the intellectual period. In his *Phenomenology of Spirit*, Hegel explains the dialectical process in which consciousness is built and transformed to become other than itself. His method consists of thinking the contradictions and superseding them through a new phase, that of synthesis. Michel Cazenave advocates a reading of Hegel's *Logic*

> [...] as it is and not as a lazy tradition has often trans-
> mitted: "For ordinary experience itself testifies that *there
> do exist* at least *a great many* contradictory things,
> contradictory dispositions etc., of which the contradic-
> tion is present not in any external reflection but right in
> them. Nor is contradiction to be taken as an abnormality
> which happens only here and there, but it is rather the
> negative in its essential determination."[71]

Cazenave locates the foundation of the primacy given to the 'work of negation' in the primordial 'nothingness' of Proclus:

> [...] In this work of negation, and perhaps we should say
> in this primacy of negation under the species of non-
> being which precedes the One and its production of
> being – primacy that even denies non-being to plunge
> into the abyss of the 'bar' – and in all the contradictions
> that arise up to the extreme of the procession, there are
> also the foundation of Hegelian logic.[72]

For Proclus, "it is more proper to reveal the incomprehensible and indefinable cause which is the One through negations; for assertions slice up reality, whereas negations tend to simplify things from distinction and definition in the direction of being uncircumscribed, and from being set apart by their proper boundaries in the direction of being unbounded."[73] Any negation must itself be negated.

'Negation of negation' is symbolized by the three, which is the stage of reconciliation or the synthesis of opposites. The idea is proposed (thesis) and then passes into its opposite (antithesis) and returns to itself by reconciling the thesis and antithesis. The synthesis is not a simple mixture of thesis and antithesis, but instead their sublation (*aufheben*), which aims to resolve their tension while overcoming adversity.

This process is very similar to what Jung in psychology called 'individuation.' It is a complex process traveling through different stages of awareness and differentiation; its purpose is to develop the individual personality. The individual is invited to align with himself and with others by confronting and integrating unconscious contents.[74]

A step in this transformation is the so-called 'transcendent' function whereby the individual must assume its opposite, the 'shadow', in transcending the conflicting oppositions. The synthesis in this case is not a simple mixture either. "The activated unconscious appears as a flurry of unleashed opposites and calls forth the attempt to reconcile them, so that, in the words of the alchemists, the great panacea, the *medicina catholica*, may be born."[75] It is a requirement of completeness for the entire psyche.

Jung points out that:

> the new personality is a third thing midway between conscious and unconscious, it is both together. Since it transcends consciousness, it can no longer be called 'ego' but must be given the name of 'Self.' Reference must be made here to the Indian idea of the Atman, whose personal and cosmic modes of being form an exact parallel to the psychological idea of the Self and the *filius philosophorum*. The Self too is both ego and non-ego, subjective and objective, individual and collective. It is the 'uniting symbol' which epitomizes the total union of the opposites.

As such and in accordance with its paradoxical nature, it can only be expressed by means of symbols.[76]

While the Hegelian dialectical process would appear to be a rational thing requiring the use of the intellect, the Jungian process of individuation is a psychological development process to be lived rather than thought. As an essentially irrational and unpredictable change, Jung's coincidence of opposites is symbolic and is located at a higher level of consciousness.

Between the two concepts, there is great affinity; a fact that Jung, two years before his death, recognizes in a letter to a correspondent:

I never really studied Hegel, I mean in his original works. There is no possibility of inferring a direct dependence but, as I said above, Hegel confesses the main trends of the unconscious and can be called a *psychologue raté*. There is of course a remarkable coincidence between tenets of Hegelian philosophy and my findings concerning the collective unconscious.[77]

Turning back to the number archetype, although Hegel also alludes to number Four, it is the Three which prevails in his 'triadic' model of the three moments of his dialectic: thesis, antithesis, and synthesis. The Three is the fundamental principle which emerges at the third term synthesis corresponding to the release of energy previously locked in a state of indecision. However, for Jung, "the Trinity is not a natural ordering pattern, but an *artificial* one."[78]

The Four is central and profoundly permeates all his work:

In an emotional opposition, i.e., in a conflict situation, thesis and antithesis cannot be viewed together at the same time [...] The unspeakable conflict posited by duality resolves itself in a fourth principle, which restores the unity of the first in its full development. The rhythm is built up in three steps but the resultant symbol is a quaternity."[79]

Illustration of the Three with the Symbolism of the Zodiac.

The symbolism of the zodiac shows the Three in the series of the three signs making up each of the four quadrants-seasons: 1. cardinal (creation, beginning), 2. fixed (inertia, hold) and 3. mutable (change to something else). For the spring quadrant, we have:

Cardinal = Aries = primordial impulse
Fixed =Taurus = stability and permanence
Mutable = Gemini = synthesis

Mutable signs have a dual appearance, illustrating the synthesis of the first two signs. Faced with the impact of the irresistible force of Aries on the immovable object of Taurus, Gemini seeks to adjust these irreconcilable forces to let emerge a new quality represented by the fourth sign of Cancer. (Fig. 10)

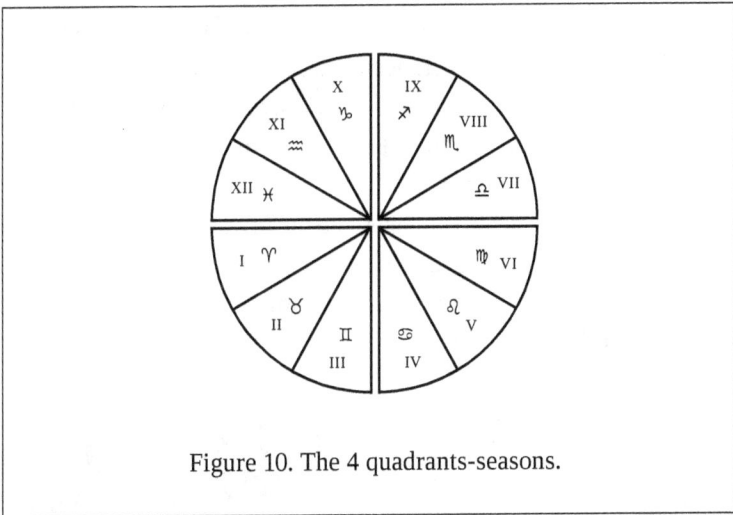

Figure 10. The 4 quadrants-seasons.

This last one can perhaps be read through the symbolism of the Four, as a return to the primordial One or as the first cardinal sign of the summer quadrant, followed by the fixed sign of Leo and the mutable sign of Virgo.

The cardinal signs of the equinoxes (Aries and Libra) are male signs that correspond to unstable equilibria leading to abrupt changes. These are the first moments of the triads 'cardinal, fixed, mutable,' which, here, are positive, centrifugal, in consonance with the dialectical process of Hegel. In contrast, female cardinal signs of the solstices (Cancer and Capricorn) correspond to states of extreme stability, closer to the 'Primitive Nature' of Schelling for which the starting point is the enveloped Being or nature as the ground of existence. Schelling observes a ternary development which, unlike Hegel, starts with a negative position. "Only the power that contracts and represses the being can be the initiating power."[80] Priority is given to the negation. "Darkness and conceal-ment are the dominant characteristics of the primordial time. All life first becomes and develops in the night; for this reason, the ancients called night the fertile mother of things."[81] As in alchemy, a degradation is necessary before reaching a fulfillment. And for anything to develop, a resistance must first be overcome. "All birth is a birth out of darkness into light: the seed must be buried in the earth and die in darkness in order that the lovelier creature of light should rise and unfold itself in the rays of the sun."[82] Schelling thus describes three moments characterized by an incessant confrontation of forces caught in an inescapable cycle[83], summarized in the third column of Table II, in correspondence with the dialectical process of Hegel and the three zodiacal qualities.

Table II

The Three Moments	Triplicities of the Zodiac	The Primal Nature or Schelling's "God's Eternal Nature"
Thesis	CARDINAL: introduction	Negating, inward-turning, contracting force.
Antithesis	FIXED: stabilization.	Affirming, outwards-flowing, expansive force
Synthesis	MUTABLE: distribution	Third force, being the unity of two, opposes and overcomes this conflict.

In summary, Three symbolizes a manifestation process that can be understood in a three-phase linear succession:[84]

1. The first phase is that of generation and expansion of a first impulse; unconscious manifestation of unity trying to persevere in its fullness.

2. The second phase is that of a contradictory tension between the One who tries to persevere in its unity, while the Other resists, wishing to differentiate from it in order to exist by itself.

3. The third phase corresponds to a superseding of contraries. There is a movement towards reconciliation and development of a dynamic, making perceptible accession to consciousness.

Four or the Tetrad

Four is the first number in the realm of manifestation. It holds a very special place in all civilization, since quaternary symbols are universally popular. Pythagoras values Four as a fundamental structure of the cosmos: four directions, four seasons, four lunar phases. And we have also four evangelists, four suits of playing cards, etc.

The quaternity, for Jung,

> [...] is an organizing schema par excellence, something like the crossed threads in a telescope. It is a system of co-ordinates that is used almost instinctively for dividing up and arranging a chaotic multiplicity, as when we divide up the visible surface of the earth, the course of the year, or a collection of individuals into groups, the phases of the moon, the temperaments, elements, alchemical colors, and so on.[85]

Illustration of Four in Zodiacal Symbolism

Zodiacal symbolism illustrates Four according to 4 elements uniting 4 triplicities (Table III)[86]

Table III

FIRE	Aries, Leo, Sagittarius	combustion, expansion, animation
EARTH	Taurus, Virgo, Capricorn	heaviness, condensation, fixation
AIR	Gemini, Libra, Aquarius	diffusion, dilation, impregnation
WATER	Cancer, Scorpio, Pisces	absorption, dissolution, fluidity

The Four Psychological Functions

For Jung, the quaternity is the archetypal foundation of the human psyche, the wholeness of the psychic processes of the conscious and the unconscious.

> In order to orient ourselves, we must have a function which ascertains that something is there (sensation); a second function which establishes *what* is (thinking); a third function which states whether it suits us or not, whether we wish to accept it or not (feeling); and a fourth function which indicates where it came from and where it is going (intuition).[87] (Fig. 11)

The goal should be to develop all four functions, but usually only one or two, the dominant and auxiliary functions are developed fastest and earliest in the course of a lifetime.

The number four, Von Franz says, "[…] always points to a totality and to a total conscious orientation, while the number three points to a dynamic flow of action. You could also say that three is a creative flow and four is the clear result of the flow when it becomes still, visible and ordered."[88] Unlike the dynamic totality of three, the wholeness of four is constituted in its interior depth.

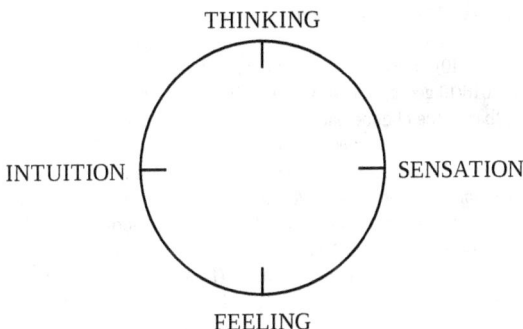

Figure 11. The 4 Jungian functions.

As the next chapter will deal entirely with the difficult step from three to four, what can be said of number five? Briefly, when the state of four is reached, the opposing forces in the Self have reached a new state of balance, symbolized by the four sides of a square. Its center, where all opposites are resolved is the *quintessence*, the fifth element of alchemy that Plato and Aristotle explained in relation to the elements and the qualities. Five is, therefore, the center of four, the center of the unfolding oneness, i.e., the "One who is not" or superessential nothingness that was to be replaced by zero when the latter found its way to Europe around the 12th century.

ENDNOTES

67   M.-L. von Franz, *Number and Time: Reflections leading toward a unification of depth psychology and physics* (Evanston, IL: Northwestern University Press, 1974), 65-6.

68   Jean-Pierre Brach, "Histoire des courants ésotériques et mystiques dans l'Europe moderne et contemporaine," in École pratique des hautes études, Section des sciences religieuses / *Vol. 106/ N°102* (1993): 338. [translated by the author] http://www.persee.fr/doc/ephe_0000-0002_1993_num_106_102_14924

69   von Franz, *Psyche and Matter*, 256.

70   von Franz, *Number and Time*, 104.

71   Michel Cazenave, "Les mathématiques et l'âme chez Proclus," 436, who quotes Hegel, G. W. F. *The Science of Logic*, George di Giovanni (ed., tr.), Book Two, (Cambridge: Cambridge University Press, 2010), §287, p. 382. Available on line at https://academiaanalitica.files.wordpress.com/2016/10/georg_wilhelm_friedrich_hegel__the_science_of_logic.pdf

72   Ibid., (The 'bar' – the oblique bar – refers to Lacan's theory of the barred Other, which designates the symbolic order as barred, incomplete, or inconsistent. The barred One corresponds to what theologians call the superessential Nothingness.) [translated by the author]

73   Proclus, *Commentary on Plato's Parmenides*, trans. Glenn R. Morrow and John M. Dillon (Princeton NJ: Princeton University Press, 1987), Book VI, §1074, p. 427.

74   Hester McFarland Solomon, "The transcendent function and Hegel's dialectical vision," in *Who Owns Jung?* ed. Ann Casement (London: Karnac, 2007), 265-90.

75   Jung, *The Practice of Psychotherapy*, C.W.16. (Princeton, NJ: Princeton University Press,1954), § 375.

76   Ibid., § 474.

77   Jung to Joseph F. Rychlak, April 27, 1959 in *Letters 1951-1961,Vol. 2*, 500-02.

78   Jung, *Psychology and Religion*, § 246, p. 167.

79   Ibid., § 258, p. 174.

80   Frederich W. J. Schelling, *The Ages of the World, Third Version (c. 1815)*, trans. Jason M. Wirth (Albany, NY: State University of New-York Press, 2000), 17.

81   Alistair Welchman & Judith Norman, "Creating the Past: Schelling's Ages of the World," in *Journal of the Philosophy of History*, vol. 4, issue 1, (2010): 29. DOI: 10.1163/187226310X490034

82   Ibid., p. 30.

83   Ibid.

84   See von Franz, *Number and Time*, p. 124, and Jung, *Psychology and Religion*, § 180, p. 118-19.

85   C. G. Jung, *Aion: Researches into the Phenomenology of the Self*, CW 9, Part II, 2nd ed., tr. R.F.C. Hull, (Princeton NJ: Princeton University Press, 1969 and London UK: Taylor & Francis Ltd, 1991), § 381.

86   Patrice Guinard, *Avatars of the Astrological Zodiac*, Available online at http://cura.free.fr/25avazod.html

87   Jung, *Psychology and Religion*, § 246, p. 167.

88   M.-L. von Franz, *Creation Myths* (Boston, MA: Shambala, 1995), 254.

6. The Transition from Three to Four

Incomplete Three. Completion and wholeness of Four. The inclusion of the feminine and the differentiation of consciousness. 'Meaning' as a new scientific worldview.

Much of Plato's *Timaeus*, which contains Trinitarian thinking, deals with the relation of Three and Four, starting with the very first words, when Socrates says in a seemingly casual way: "ONE, two, three; but where, my dear Timaeus, is the fourth of those who were yesterday my guests and are to be my entertainers today?"[89]

Jung, who considers trinity as an incomplete quaternity requiring completion by adding a fourth realizing principle, is interested in the appearance of quaternity among philosophers and theologians. Besides Plato's *Timaeus*, he cites Goethe, who, in the Cabiri scene makes Faust declare: *"Three along we've brought, But come the fourth would not, He said, he was the right one Who thought for all of them."*[90]

For Jung, the fourth is Goethe's own thinking function:

The Cabiri are, in fact, the mysterious creative powers, the gnomes who work under the earth, i.e., below the threshold of consciousness, in order to supply us with 'lucky ideas.' As imps and hobgoblins, however, they also play all sorts of nasty tricks, keeping back names and dates that were 'on the tip of the tongue,' making us say the wrong thing, etc. They give an eye to everything that has not already been anticipated by the conscious mind and the functions at its disposal. As these functions can be used consciously only because they are adapted, it follows that the unconscious, autonomous function is not or cannot be used consciously because it is unadapted. The differentiated and differentiable functions are much easier to cope with, and (for understandable reasons) we prefer to leave the 'inferior' function around the corner, or to repress it altogether, because it is such an awkward customer. And it is a fact that it has the strongest tendency to be infantile, banal, primitive, and archaic. Anyone who has a high opinion of himself will do well to guard against letting it make a fool of him. On the other hand, deeper insight will show that the primitive and archaic qualities of the inferior function conceal all sorts of significant relationships and symbolical meanings, and instead of laughing off the Cabiri as ridiculous Tom Thumbs he may begin to suspect that they are a treasure-house of hidden wisdom. Just as, in *Faust*, the fourth thinks for them all, so the whereabouts of the eighth should be asked 'on Olympus.' Goethe showed great insight in not underestimating his inferior function, thinking, although it was in the hands of the Cabiri and was undoubtedly mythological and archaic. He characterizes it perfectly in the line: 'The fourth would not come.' Exactly! It wanted for some reason to stay behind or below.[91]

Jung's concern about the fourth is connected to his conception of the transcendent function that transcends the conflicting oppositions while seeking union of consciousness and the unconscious. This had been intuitively understood by alchemists and Jung, very often cites the axiom of Maria Prophetissa, which is a sort of alchemical analogy of the process of individuation. It shows the concrete result of the binary and ternary deployment from unity, with the idea of a return to a whole: "One becomes two, two becomes three, and out of the third comes the One as the fourth."[92]

6.1. The Transition from Three to Four in the (3 + 1) Psychological Functions

Jung illustrates the step from Three to Four in the phases of consciousness of an individual in terms of the development of the fourth psychological function. One of these faculties dominates and determines our psychological type. If thought dominates (see Fig. 11), its opposite, feeling, is repressed into the unconscious and vice versa. If intuition dominates, sensation is repressed and vice versa.

To achieve fullness, we must look for the repressed faculty in the unconscious and integrate it with the three other conscious faculties. The 'lower' function coincides with the dark side of human personality. Its integration into consciousness corresponds to the passage from Three to Four, allowing a return to the One, not through a regression to the fusional state but in a fully differentiated plane of consciousness.

6.2. The Transition from Three to Four in the Trinity and the Assumption of Mary

Jung was quite interested in the Assumption dogma of Mary of the Catholic religion. In his view, this recognition dating from 1950, had actualized a presence already unconsciously felt by the masses in the form of the worship of Mary during many centuries. (Fig. 12)[93]

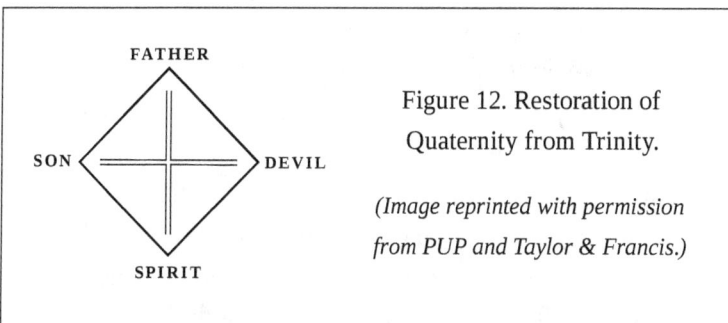

Figure 12. Restoration of Quaternity from Trinity.

(Image reprinted with permission from PUP and Taylor & Francis.)

The inclusion of the feminine, symbolized by the Virgin Mary in the male abstract structure of the Trinity, symbolizes a progression in consciousness. In fact, according to Jung, the fourth principle can have forms as varied as the principle of evil, the devil, the earth, matter, the body and the feminine. Evil, Jung insisted, is a force in itself. The feminine is supposed to fight against evil. Genesis 3:15 says: "I will put enmities between thee [the serpent] and the woman, and thy seed and *her* seed: *she* shall crush thy head, and thou shalt lie in wait for *her* heel [emphasis added]." Mary (Four), being the new Eve (Two) who disobeyed God is opposed to the dark evil. Thus, Mary and the devil form a conjunction of opposites and Jung writes: "These two incompatible figures are united in the Mercurius duplex of alchemy."[94]

6.3. (3 + 1) Dimensions of Space-Time and (3 + 1) Interactions in Physics

In a quaternity, one of the four terms often occupies an exceptional position or has a different nature from that of the others. For example, in the representation of the evangelists, three of the four symbols are animals, and the fourth is an angel. In modern physics, the three spatial dimensions (length, width, and depth) form a single, indivisible entity with the temporal dimension.

Thus, we can see space-time as a (provisional) synthesis marking the progress of consciousness. Let us note that a space-time which would consist of the three modes of time – past, present, future – and a spatial dimension would also form a (3 + 1) structure.

The four known forces or interactions of physics (weak nuclear, strong nuclear, gravitational, and electromagnetic) also form a (3 + 1) structure, not yet melded into a single physical theory. The force of gravity is uniquely positioned due to its extreme relative weakness, which makes synthesis extremely difficult. The other three forces have already theoretically been linked and the search for a unified theory, combining the four, continues.

In terms of structure formations in the universe (galaxies, atoms, animals, etc.), the weak nuclear force is not responsible for any structure. It is thus distinguished from the "triad of active entities of the universe:"[95] gravitational, strong nuclear, and electromagnetic.

6.4. The Transition from Three to Four in the (3 + 1) Quantum Numbers

Pauli was a very important actor in the construction of the four quantum numbers theory. Until 1925, spectral lines had been explained combining three numbers that characterized size, angle and shape of the electron orbit inside an atom. It turned out that four quantum numbers were necessary to characterize the state of the electrons – in contrast with three degrees of freedom sufficient to define the classical motion of electrons in a three-dimensional space. What was the physical meaning of the fourth quantum number? If the electron rotation was affected in the manner of Earth rotating on its axis, there would be *no need* of a fourth number, because in classical physics such a movement is perfectly described.

In quantum physics, this new number – the 'spin' – had to have two possible states that could 'classically' be represented as an electron orbiting the nucleus and rotating on a vertical axis, either clockwise or anti-clockwise. This fourth number, different in nature from the other three quantum numbers, led Pauli in 1925 to the intuition of one of the great rules of physics, which is the Exclusion Principle: electrons in an atom *all* have different quantum numbers.

While commenting on his discovery, he recognized that he had been inspired by the transition from Three to Four: "[...] My path to the Exclusion Principle had to do precisely with the difficult transition from 3 to 4, namely with the necessity to attribute to the electron a fourth degree of freedom (soon explained as 'spin') beyond the three translations ... That was really the *main work*."[96] In his mind, this discovery reflected his own psychological journey that had led him to carry out the transition of the three psychological functions to the fourth. In his personal case, this meant integrating into consciousness the repressed 'feeling' function in the unconscious.

Theoretically, this rule, that Pauli had extracted from the World Soul, was later explained by Fermi-Dirac statistics (1926), and Pauli himself participated in the extension of the theory. Ultimately, all particles belong to either one or the other of the following categories: fermions, *obeying* the exclusion principle, are such that the 'state vector' is antisymmetric when exchanging two identical particles; and bosons, *not obeying* the exclusion principle, for which the 'state vector' is symmetric while exchanging two identical particles.

6.5. Trinitarian Kepler and Quaternarian Fludd

Interested in the influence of archetypal representations on the formation of the scientific theories of Johannes Kepler, Pauli could not ignore the famous controversy which took place between the inventor of the three laws of celestial mechanics and Robert Fludd, his physician and alchemist contemporary, close to Paracelsus' ideas. Kepler and Fludd were both steeped in mysticism and followers of the Pythagorean tradition of number, yet they were opposed to what could be summed up as heartbreak between Three and Four.

Pauli admired astrologer-astronomer Kepler for having seized, through the intuition and mediation of numbers, the nature of cosmic order and for having given impetus to physics: "his ideas represent a remarkable intermediary stage between the earlier, magical-symbolical and the modern, quantitative mathematical descriptions of nature."[97]

Nevertheless, he noted that symbols did not contain reference to quaternity. They were dominated by the archetype of Three, such as the Christian Trinity that showed a cosmos where the Earth revolved around the Sun, with God at the center of a sphere from which light emanated. The Earth was the creation of God – as his Son – and the eternal orbiting movement was the Holy Spirit "in the equality of relation between point and circumference." [98]

Pauli admired just as much the alchemist Fludd, who remained attached to a *qualitative* explanation based on symbols while trying to preserve the unity of inner experience and of external processes of nature. Its symbols, close to oriental mandalas, are dominated by the archetype of Four. This is where Pauli notes an important indication in the complaint that Fludd addresses to Kepler: "[...] you force me to defend the dignity of the quaternity (*cogis me ad defendam dignitatem quaternarii*)."[99]

Pauli knew very well that in a very short time the emerging scientific approach, initiated by Kepler, would distance the simple drive to know and, with the impulses of Galileo, Descartes, and Newton, within a few decades, science would lead to a mechanistic approach of nature where the observer would be separated from the physical processes.

It's an approach to science that Pauli cannot accept and out of which he perceives lineaments in Kepler's work, particularly in the will of the latter to make astrology scientific, a position which he seems to mock. Pauli well recognizes in Kepler a certain archetypal approach to astrology as an inspiring science of symbolic forms, but he cannot follow him when he attempts a causal explanation based on light rays coming from celestial bodies that would have astrological effects on individuals.

Consequently, Pauli explains that,

> if one proceeds on this basis it hardly appears possible to avoid the empirically untenable conclusion that artificial sources of light would also be able to produce astrological effects. In general, I should like to remark in criticism of

astrology that, in consequence of the vague character of its pronouncements (including the famous horoscope that *Kepler* drew up for *Wallenstein*), I see no reason to concede to horoscopes any objective significance independent of the subjective psychology of the astrologer.[100]

In Pauli's eyes, the mutual Kepler-and-Fludd incomprehension – similar to, later in time, Goethe's criticism against Newton's theory of colors – lies in the fact that, obviously, none of them is speaking 'at the same level' of reality. Fludd, like Goethe when speaking of a qualitative unity of the world, is dealing with a metaphysical and spiritual realm; Kepler, as Newton when talking about quantities and purely external features of nature, is in a scientific realm.

For Pauli, Fludd's quaternary approach supported a qualitative complementarity to the scientific quantitative approach of Trinitarian Kepler. While considering the role of the observer in the observed phenomenon, quantum physics has approached the quaternary perspective. The Kepler-Fludd controversy that Pauli tracks in the history of ideas is but a reflection of the difficult (3 + 1) transitions that dot the historical development of scientific ideas.

It is the same qualitative complementarity that led to agreement between Jung and Pauli on promoting the notion of 'meaning' in a new scientific worldview, recognizing the different levels of reality. Meaning emerges in observed a-causal phenomena *in* subatomic physics and, *in* psychology, in the spontaneous link between the physical sphere and the psychic sphere, inner and outer worlds. But the "realization of 'meaning' is [...] not a simple acquisition of information or of knowledge, but rather a living experience that touches the heart just as much as the mind. It seems to us to be an illumination characterized by great clarity as well as something ineffable – a lightning flash ..."[101] For Jung and Pauli, recognizing the existence of meaning autonomous with respect to the contemporary consciousness is somehow a transdisciplinary step towards Fludd. In search of the stage from three to four, they will come to a global vision of quaternary form, where synchronicity appears as the missing dimension to integrate.[102] (Fig. 13)

"Meaning," Jung wrote to Erich Neumann March 10, 1959, "is always unconscious and can only be discovered post hoc; this is why the danger also always exists that meaning will be insinuated where nothing of the sort is present. We do need the synchronistic experiences to be able to justify the hypothesis of a latent meaning that is independent of consciousness."[103]

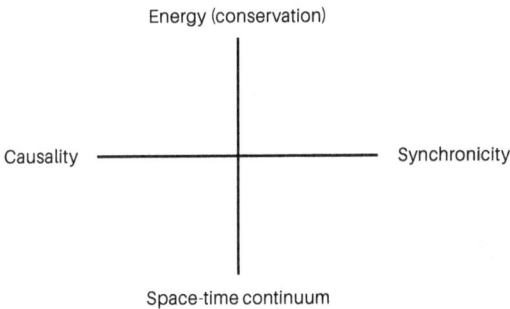

Energy (conservation)

Causality ——————————————— Synchronicity

Space-time continuum

Figure 13. Quaternio encompassing synchronicity.

ENDNOTES

89   Plato, *Timaeus*. Trans by Benjamin Jowett, Ed. By E. Hamilton and H. Cairns. Collected Dialogues of Plato. (Princeton NJ: Princeton University Press, 1989).

90   Jung, *Psychology and Religion*, § 243, p. 164.

91   Ibid., § 244, p. 164.

92   C. G. Jung, *Psychology and Alchemy, CW 12*, tr. R.F.C. Hull, (Princeton, NJ: Princeton University Press, 1980 and London, UK: Taylor & Francis Ltd, 1989), § 26.

93   Jung, *Psychology and Religion*, § 258, p. 175.

94   Jung, *Aion*, § 397, p. 253.

95   In Tao, the Heaven-Earth-Man triad of "Active Entities of the Universe" combines with the four seasons Tetrad to form 4^3 = 64 possibilities of becoming. See: Michel Vinogradoff, *L'Esprit de l'Aiguille – l'apport du Yi Jing à la pratique de l'acupuncture.* (Paris: Springer-Verlag, 2010).

96   Arthur Miller, *137: Jung, Pauli, and the Pursuit of a Scientific Obsession*, (New York, NY: W.W. Norton & Company, 2010), 66.

97   Pauli, "The Influence of Archetypal Ideas," in *Writings*, 222.

98   Ibid., 225, who quotes "Kepler's Mysterium Cosmographicum."

99   Meier, ed., *The Pauli/Jung Letters,* appendix 6, 208.

100   Pauli, ibid., 243.

101   von Franz, *Psyche and Matter*, 257.

102   Pauli to Jung, November 24, 1950, in C.A. Meier, ed., *Atom and Archetype*, no. 45P, p. 57.

103   Jung to Erich Neumann, March 10, 1959, *Letters 1951-1961, Vol. 2*, 495.

7. Traces of a Quaternary Rhythm in Cosmological Events

Four as self-emerging organizational qualities.
Re-union of the two 'scars' of time as the Fourth.
The 'undifferentiated' of contemporary research
in physics. Four Aristotelian causes and Four-step
alchemical work.

The evolution of physics has always depended on cultural and religious contexts. Its very emergence presupposes, beforehand, a linear model of time. Represented by an arrow oriented from past to future, this continuous time – irreversible and without end – is a relatively recent notion that appears with history. Succeeding cyclical visions of renewal, it emerges with major cosmologies which take up and extend cosmological myths.

With Judaism and then Christianity, linear time breaks with the eternal return of 'archaic' humanity. In contrast to the Greek view, the Christian conception of time starts with the first chapter of *Genesis* and ends in the eschatological perspective of the *Apocalypse*. It then becomes an arrow line with a beginning and is punctuated with events that respect the principle of causality: every event is the effect of a cause that preceded it. For Newton, time is common to all observers, identical in all places. It flows evenly, and it is represented by the one-dimension mathematical variable t. Absolute, it allows us to date all events wherever they take place.

But with Einstein's relativity, space and time become relative, which is to say that they *depend* on the observer. As separate quantities, "space has no objective reality except as an order or arrangement of the objects we perceive in it, and time has no independent existence apart from the order of events by which we measure it."[104] There is no more meaning in a statement of the time of an event in the universe. Space and time become the 4-dimensional *space-time* that Einstein identifies with the universe, hence his first model.

However, the currently dominant model of the universe is based on the cosmological principle of spatial homogeneity and isotropy which is well verified by observations. This very principle allows us to simplify the relativistic space-time and define a *unique* cosmological time t, related to universe expansion. Metric is simplified, and spatial distances depend on a scale factor $R(t)$.

Cosmological time shares with Newton's absolute time the property of being universal. It is thus an arrowed time, graduated from zero to infinity, allowing for the narrative of the universe's history while dating events. However, theoretical 'zero' time, which would reflect an infinite temperature, does not exist. Physics is powerless to describe phenomena for temperatures exceeding a limit of 10^{32} degrees, called 'Planck temperature'. Time, thus, does not begin at time zero, but at the end of the 'Planck era' or 10^{-43} seconds (a hundred billionth of a billionth of a billionth of a billionth of a billionth of a second) after zero. (Fig. 14)

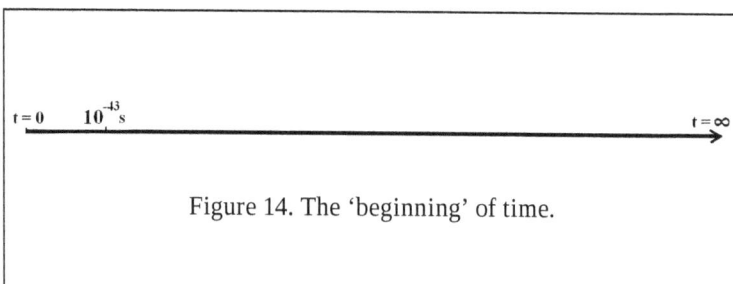

Figure 14. The 'beginning' of time.

The cosmological narrative is ordered on this time axis. As far as it is a story, is it possible, as in ancient cosmologies, to identify archetypal and mythological dimensions? In fact, events that mark the time axis are not homogeneous, because all points of time do not have the same amount of information. Archetypal traces reflecting the quality of numbers might emerge in the temporality of the narrative as qualitative events or rhythms that appear to escape the temporal succession law. Reflections of ordering factors, common to matter and psyche, could be revealed in transitions or breaks in the course of events, and beginning with the par excellence 'imperfection' represented by the 'no-beginning' of time.

In the depths of the unconscious, time remains cyclical. It is the thrust of consciousness that operates the break of the time circle, whose symbolic marks on the cosmological temporal axis are 'zero' and 'infinite.' The cut carries with it psychological qualities of One and Two which, as von Franz writes in her book *Creation Myths*, appear as two fundamental numbers that cannot exist without each other and illustrate the struggle between an original diffuse and the individuation of a will.[105]

She notes that if Three frequently appears, Four appears all the more frequently. Equally, according to Edward Edinger: "[...] the division process into four is a primary cosmological image."[106] In particular, in Judaism and Christianity, the cosmological picture is described in the four great archetypes of *Genesis*.

If there were to be archetypal dimensions in the modern cosmological narrative, a possible approach is to look for the flow of events spread over a time axis, in the chronology of their linear scheduling. What could and may be prioritized are self-emerging organizational qualities combined with qualitative notions borrowed from statistical mechanics – such as attractor, horizon, symmetry breakings and phase transitions, as well as any reflections of rhythmic configurations of energy.

The large-scale structure of the universe is only revealed by ignoring the stars of the Milky Way and nearby galaxies. Likewise, an overall underlying configuration to chronological events will only be revealed by bracketing those events that relate in terms of cause and effect; among them, the formation of atomic nuclei in primordial nucleosynthesis, that of light atoms (hydrogen, helium, and lithium) or that of heavier elements in stars, or still, the formation and evaporation of black holes, which are events planned for the far future of the universe.

Of course, causes and effects are not necessarily mapped linearly. But the point here is to sort out events that break temporal continuity. They may refer to the *strong emergence* concept which qualifies incomprehensible processes from the point of view of causal laws and forces. Some other events, instead of a temporal break, seem to *freeze* time and are associated with notions of horizon or time attractors.

7.1. Two *Strongly Emergent* Events

A macroscopic property is labeled *strongly emergent* if it tends to act in a causally-oriented manner, from whole to component parts. Levels of more complex events thus affect less complex events. *Strong emergence* differs from *weak emergence*, in which there is a 'bottom-up' causal link determinable in the manner of 'Laplace's demon,' starting from simple to complex.

This latter category comprises phenomena which have causes that science could (or will be able to) explain reductively. Such is the case of water, whose liquid state seemed to emerge from properties of hydrogen and oxygen gases, a property today explained following progress in knowledge of the chemical elements that compose it. Similarly, superconductivity, an emergent phenomenon discovered in 1911, has since been causally explained.

In all cases of *weak emergence*, one prefers to describe the system by its global descriptive properties rather than explain it by a long chain of reason: why, for example, a die falls on this or that side. In the history of the universe, the emergence of life is a *weakly emergent* phenomenon among many others involved as the universe cools, punctuated by breaking symmetries that reveal increasingly differentiated structures.

But two events stand out from the story by their *strong emergence*. They have in common that they both been 'added' to the big picture of the evolution of the universe. Indeed, the emergence of consciousness still appears as an epiphenomenon, most of the time ignored by cosmologists. Similarly, the theory of cosmic inflation – which presently appears unavoidable – is an *ad hoc* theory added to the cosmological fresco, initially motivated by the need to explain alleged 'paradoxes' like those of flatness and near homogeneity at the large scale of the fossil radiation.

1) First *strongly emergent* event: Inflation, developed to bring the universe out of the fluctuating quantum vacuum, corresponds to a *strongly emergent* phenomenon of the early universe, creating space and time fitted with its arrow at the age of 10^{-35} seconds. (Fig. 15) By involving a hypothetical scalar field called the 'inflaton,' whose strength increases exponentially, some physicists attempt to explain the extraction of the huge vacuum energy – whose density is that of Planck conditions – towards a stabilized phase where the scalar field becomes a field of particles. Although the theory does not offer a clear shut-off mechanism, this process is assumed to stop at ~10^{-32}s. As a *strongly emergent* phenomenon, cosmic inflation questions the relation of wholeness to parts and the notion of downward causation; namely, the idea that an entity located at one ontological level may have a causal influence on an entity located on a lower ontological level. Here, what would have causally created inflation is lost in the quantum vacuum entanglement defined in physics as the ground state.

Other physical approaches, assuming a quantification of space-time, suggest a link of inflation with some form of consciousness. Computational Loop Quantum Gravity postulates that the universe had a moment of conscious experience at the end of inflation when it was selected out of a superposition of many by a self-reduction mechanism. It is termed as the 'Big Wow' by Paola Zizzi[107] who, drawing on the work of Whitehead, Chalmers, and others, as well as Penrose and Hameroff, explains how consciousness arose in the universe and "acted to indelibly imprint on future minds to come, dictating future modes of computation, consciousness and logic."[108]

2) **Second** *strongly emergent* **event:** Cosmology is obliged to recognize that the observer is part of its object and that, somehow, this observer emerged with *Homo sapiens* (at about 300,000 years of margin) at an age of 13.8 billion years. (Fig. 15) This is a second major discontinuity. The emergence of consciousness makes the universe pass a new step, that of reflection. Through *Homo sapiens,* the universe knows that it knows. In some theories, this 'strongly emergent' phenomenon has been associated with a spontaneous biological and chemical-physical breaking of symmetry in the structuration of the cerebral system which is the most complex object of the universe.

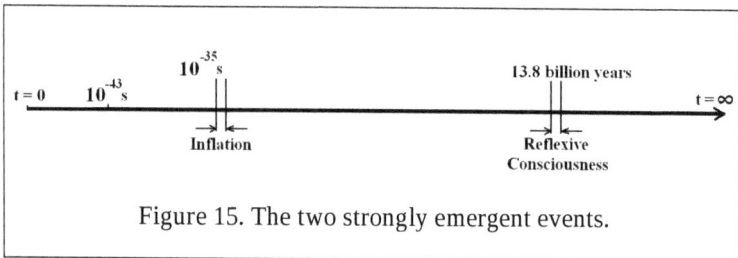

Figure 15. The two strongly emergent events.

In one of these theories, the brain is considered as an open system consisting of 'condensed matter,' in continuous interaction with the environment. The dominant symmetrical component in living matter is the electric dipole of water molecules. The stimuli of the outside world may trigger the symmetry breaking of the brain's microscopic dynamics with the result of producing recoding memory. In his *dissipative quantum model,* Giuseppe Vitiello discusses the 'brain / environment' system as a closed system 'brain and its Double.'

Consciousness then belongs to the bridge which connects, does not separate the self from its Double. It lives in the present since the present stays on the surface of the mirror in time in which the self reflects in its Double, and vice-versa. It is interesting to observe that the word συνειδώς, which means to 'see together,' in the act of 'immediate vision,' was used by the ancient Greeks to denote consciousness (to be conscious of), thus stressing the 'present' as the time dimension of consciousness (the verb οράω is used instead for the act of lasting vision. Consciousness is thus an act of sudden knowledge, an intuitive one, an *unum* not susceptible to be divided into rational steps, thinkable but «non-computational», as the present is, and it is non-separable from our body.[109]

Another quantum approach to the reflexive phenomenon of consciousness is the *theory of Orchestrated Objective Reduction* (Orch OR theory).[110] Building on Gödel's incompleteness theorem, Roger Penrose believes consciousness is not a simple calculation process because computability theory prohibits generation of a conscious human intelligence by a sufficiently complex machine. He claims that the phenomenon inherent to quantum measurement – wave function collapse – provides this non-algorithmic element. Consciousness does not cause wave function collapse (as in von Neumann-Wigner and original Copenhagen interpretations), but rather it is the wave function collapse which creates 'moments of conscious awareness' or 'qualia.' Wave function collapse (or *reduction*) is equivalent to decoherence, the process by which a quantum system is disrupted by thermal interactions of its random environment. *Objective Reduction (OR)* is one of many collapse theories that Roger Penrose, in collaboration with anesthesiologist Stuart Hameroff, developed into a consistent, falsifiable hypothesis of consciousness supposed to take place within some protein polymers called 'microtubules' located in the neurons of the brain. Triggered every time that a gravitational potential energy threshold is reached, this is another example of downward causation that characterizes *strong emergence.*

7.2. Two *Horizon-*Type Events or *Temporal-Attracting* Events

Other events stand out clearly from the background of the cosmological timeline, no longer by the nature of their emergence but by their temporal 'stretching.'

1) First *horizon*-type event or *temporal-attracting* event: This concerns the opaque / transparent transition when the universe was 380,000 years old. Before, the temperature was such that most of the photons of the universe interacted with electrons and protons. Light was constantly absorbed and re-emitted, and the ionized plasma which then made up the universe was 'attracted' by the final conditions of the irreversible evolution towards 'heat death;' hence, the name *horizon* or *temporal attractor.* After, the 'recombination' released photons and 'decoupled' (visible) matter from light, allowing the universe to become transparent and locally 'fall' into gravity,[111] thus, reinforcing the name of attractor that was given to this event. (Fig. 16)

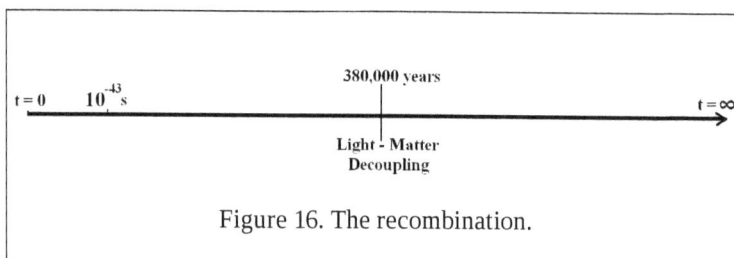

Figure 16. The recombination.

2) Second *horizon*-type event or *temporal-attracting* event: This second event which in fact, is the 'par excellence,' limit-event, combines in one feature of 'unattainable time' the two ends of the time axis.

They appear like 'scars' in linear time or like two surviving testimonies of the cut of the circle of time related to the cyclical time of the unconscious. Here, the horizon is not a phenomenological one, but it points to an extreme zone of physical theory founded on linear time. It is interesting to note that while approaching these two time-limits of cosmological time, modern physics deals with either very small or infinitely large durations; which brings to mind the concept of psychobiological time.

The linear time scale to which we are accustomed in no way reflects the pace of events in the universe as a whole. René Guénon thinks:

> It is evident that periods of time are qualitatively differ-
> entiated by the events unfolded within them, just as the
> parts of space are differentiated by the bodies they
> contain; it is not therefore in any way justifiable to regard
> as being really equivalent durations of time that are
> quantitatively equal when they are filled by totally
> different sequences of events; it is indeed a matter of
> current observation that quantitative equality disappears
> completely from the mental appreciation of duration in
> the face of qualitative difference. [...] Time is not some-
> thing that unrolls itself uniformly, so that the practice
> of representing it geometrically by a straight line, usual
> among modern mathematicians, conveys an idea of time
> that is wholly falsified by over-simplification.[112]

For Jean-Marc Lévy-Leblond, the *Big Bang* becomes an optical illusion similar to that of the temperature of 'absolute zero' or the unsurpassable character of the speed of light.[113] These numerically finite limit-values are conceptually infinite, as seen from their asymptotic character, which continuously makes the explored time-scale reduced as we approach time $t = 0$. Lévy-Leblond has defined an 'additive parameter' θ, such that as t approaches zero, his θ–time stretches infinitely far back to 'minus infinity.' Thus, instead of a creation at $t = 0$, there is eternal existence and therefore no singular beginning.

This 'initial' instant is not in time and cannot be considered as the beginning of time. It can be considered as an origin for counting time, provided one does not forget that this origin does not belong to the succession of moments laying ahead of it. This time $t = -\infty$ has become an infinite horizon exactly as time $t = +\infty$ located in the far future of the expansion of the universe where matter will eventually decay and be turned to light – which removes any physical meaning to a space without matter that would continue to expand to infinity. (Fig. 17)

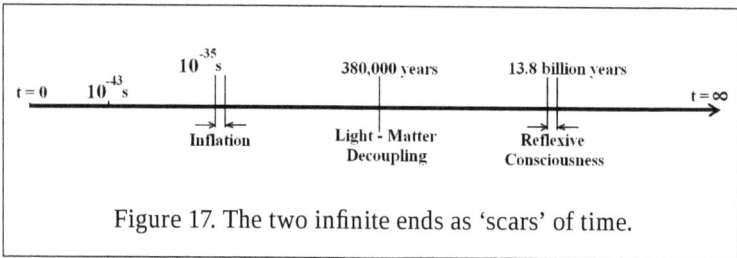

Figure 17. The two infinite ends as 'scars' of time.

In the case of a contraction into a final singularity, the *Big Crunch* is analogous to the black hole horizon, which, there too, gives an infinitely stretching time. Time depends on the gravitational field, going a bit faster in the attic than in the basement. On the surface of a neutron star it 'flows' 70% slower than on Earth and on the surface of a black hole it becomes infinitely slow. By crossing its horizon, it would be possible to see the progress of all past and future history of the universe in a single glance.

Thus, at both ends $t = 0$ and $t = \infty$ of the time axis, two *temporal horizons* appear that are both infinite. These two infinities mean that something is wrong in theory, or that we leave the framework of a theory that was made only for a certain scale. They hide what is to be discovered, waiting for a new theory. (Fig. 18)

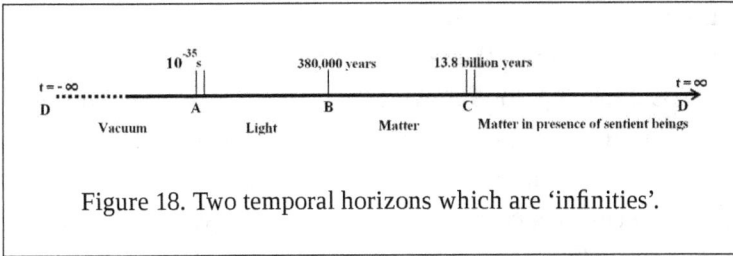

$$t = -\infty \qquad\qquad t = \infty$$

| 10^{-35} s | 380,000 years | 13.8 billion years |

D   A   B   C   D

Vacuum   Light   Matter   Matter in presence of sentient beings

Figure 18. Two temporal horizons which are 'infinities'.

The cut of cyclical time, which is the basis of the meteoric rise of science, now returns as a boomerang through its two scars (t = - ∞ and t = ∞) not fully healed at both ends of linear time: initial and/ or final singularities, reflecting the current boundaries of theoretical physics. Connecting these two points in a point D on a circle diagram may make sense for all cosmological models, and not just as the unattainable bounce limit-point of the universe during 'Big Crunches' for cyclical models. (Fig. 19)

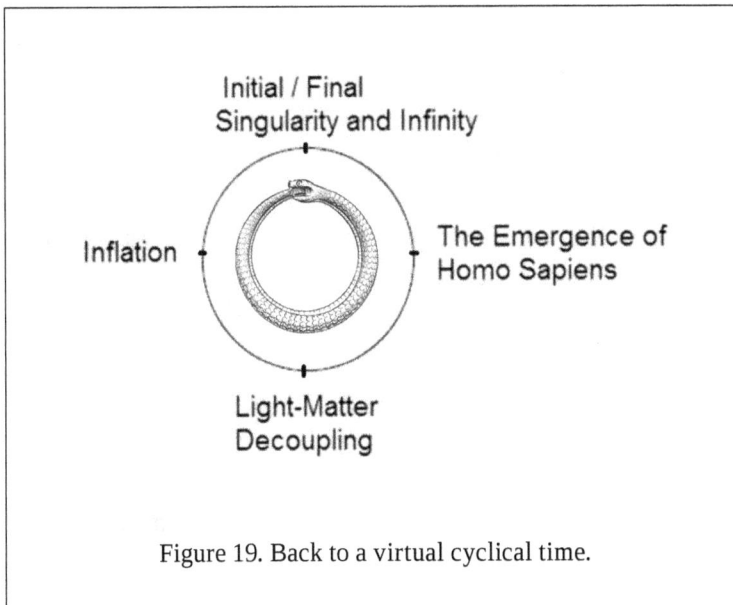

Initial / Final
Singularity and Infinity

Inflation

The Emergence of
Homo Sapiens

Light-Matter
Decoupling

Figure 19. Back to a virtual cyclical time.

This point can represent both the theoretical 'starting point' of the Plank era as well as theoretical 'end points' of the 'Big Crunch' models or asymptotic futures of post-material open universes. It can be a purely representative boundary that attaches and separates the vanishing final spacetime of an expanding universe to the next aeon in Penrose's conformal cyclic cosmology.[114] In any case, it points towards 'beyond-time' or towards a more integrating virtual time, which is that of cyclical time.

Thus, the D-point event of fig. 18 incorporates the quality of the 'fourth' in the very sense of the 'undifferentiated' of contemporary research in physics. With the other three A, B, and C previously pointed events, it may well be a left behind trace of an organizing principle likely to call for new mathematical intelligibilities.

The circular or spherical representation thus allows for the subsuming of several forms of time, including physical time and the time of alchemists for whom the beginning of The Great Work was like the end. The *opus circulare* was represented as an image in the form of a *ouroboros*, the serpent eating its own tail. For physics, reclosing the linear time axis as the simplest and most perfect geometrical figures should not appear as an unnecessary step backward, or at its worst, as a regressive operation. This could be a temporary phase of projective development, allowing for a different way of thinking the problem of time, and possibly opening new avenues for science to supersede itself.

We can explore the universe following many routes. As Alexandre Koyré wrote: "The *itinerarium mentis in veritatem* [road to truth] is not given in advance, and the mind does not proceed along it in a straight line. The road to truth is full of traps, strewn with errors, and the failures there are more frequent than the success ... But we would be wrong to neglect the study of errors – it is by way of them that the mind progresses towards the truth."[115]

7.3. Relative Durations of the Four Quadrants of the History of the Universe

Marked in terms of linear time, the four quadrants cut in the universe's history seem to have very unequal durations, as shown in Table IV. The second quadrant lasts much longer than the first one (with a ratio of 36,000). This is the previously mentioned effect of the linear time scale that does not reflect well the pace of events in the universe.

Table IV

	Fourth Quadrant	First Quadrant	Second Quadrant	Third Quadrant
	From 0 to 10^{-35} seconds	From 10^{-35} seconds to 380,000 years	From 380,000 to 13.8 billion years	From 13.8 billion years to $+\infty$
Duration in linear time	Duration = 10^{-35} seconds	Duration = 380,000 years	Duration = 13.8 billion years	Duration = ∞ years

By choosing additive time, which is the time scale proposed by Levy-Leblond[116], the durations of the quadrants I and II are approaching (Table V). In fact, it even appears that the first quadrant has the longest duration (but here in a ratio around 7).

Table V

	Fourth Quadrant	First Quadrant	Second Quadrant	Third Quadrant
	From $-\infty$ to -9.7×10^{11} years	From -9.7×10^{11} years to -1.2×10^{11} years	From -1.2×10^{11} years to 0 (Now)	From 0 (Now) to $+\infty$
Duration in additive time	Duration = ∞ years	Duration = 8.5×10^{11} years	Duration = 1.2×10^{11} years	Duration = ∞ years

But the event associated with D point continues in infinite time and, consequently, the two quadrants that surround it have infinite duration. In any case, the 'additive' time scale shows, for quadrants I and II, a more regular event distribution than in linear time. Through this scale, the universe approaches a living being. The perception of its events is similar to that of the subjective perception of the human being whose notion of the passage of time is logarithmic as postulated by Fechner.[117] His time passes more quickly as he ages.

Still with a logarithmic scale, but in a quite different perspective, Rodney Collins accounts for the difference between the value of an hour of childhood and an hour of old age. For units of biological time, he uses the lunar month (28 days). The nine solar months of pregnancy correspond, therefore, to ten lunar months: "Man is born ten lunar months after his conception; and his childhood is generally accepted as coming to an end after one hundred (7 years). [...] The full span of man's life is about one thousand such months (76 years)."[118] This progression is logarithmic (1, 10, 100, 1000) rather than linear.

What counts is the subjective time or lived time, that which consciousness perceives; namely, the ability to process information. In the same vein, physicists who have considered the persistence of life in the far future of the universe – which will be discussed later – have shown that the time associated with living species depends on the speed of their metabolism, itself being closely related to temperature.

7.4. The Four Quadrants and Cosmological 'Eras'

Gradually, as the universe expands, the number of particles of matter and photons per unit volume decreases. However, the photon energy is further reduced as a result of the cosmological redshift, which reduces their equivalent mass and hence, their density. Accordingly, upon expansion of the universe, the radiation density decreases faster than the density of matter.

In terms of energy density, the early universe is thus marked by the predominance of radiation that dominates matter before the crossing point, which is around 72,000 years old. (Chart 20)[119] Matter then dominates until the crossing point with dark energy – which has a constant energy density insensitive to expansion – at around 10 billion years.

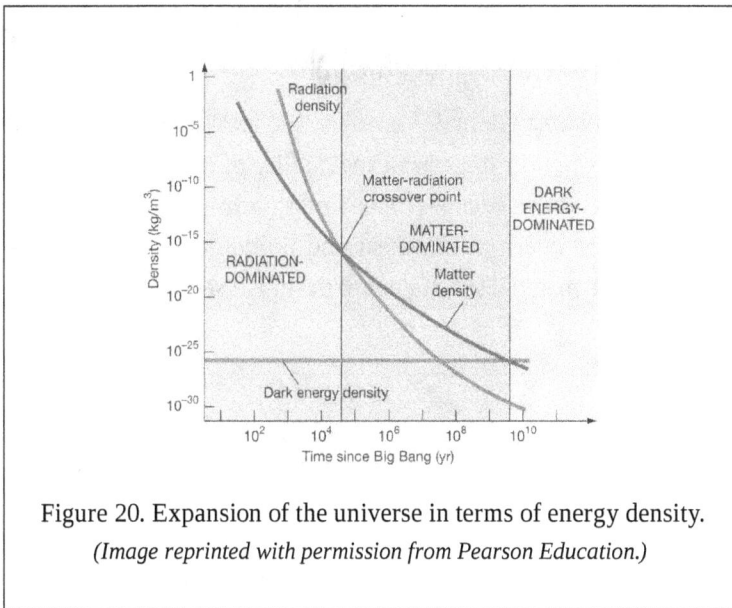

Figure 20. Expansion of the universe in terms of energy density.
(Image reprinted with permission from Pearson Education.)

There seems to be a diffuse correspondence between these three cosmological eras and the first three quadrants, the fourth quadrant escaping from this time marking. The points of intersection between first/second and second/third quadrants do not match exactly, but the overlap differences remain, however, relatively moderate: ~72,000 years instead of 380,000 years (due to preeminence of dark matter) and ~10 billion instead of 13.8 billion years. (due to preeminence of dark energy)

The modern 'energy density' concept is a distant heir of the fire element, endowed with consciousness and privileged in Heraclitean cosmology. Aristotle, meanwhile, will not grant it special prominence, relegating fire to the status of a principle, equal to earth, water, and air. On the other hand, cosmological precedence returns to its four causes, reconsidered in correspondence with the (3 + 1) cosmological eras in the following paragraph.

7.5. The (3 + 1) Aristotelian Causes and the (3 + 1) Steps of Alchemical Work throughout the Universe

As cosmic history gradually unfolds, the four causes seem to be actualized as manifestations of the quality of energy:

- **material cause** emerges in the **first quadrant**. Formless matter, resulting from quark confinement and primordial nucleosynthesis, represents the material substratum of the universe. (Fig. 21)

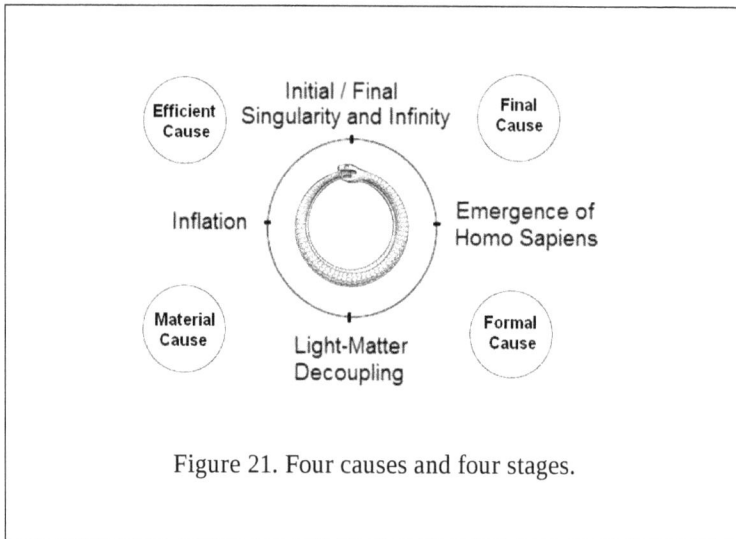

Figure 21. Four causes and four stages.

- **formal cause** is manifested in the **second quadrant** when matter, released from light, can fall into gravity and express itself in many forms. In fact, it is only from matter-light decoupling that the universe's texture is no longer homogeneous. Multiple density inhomogeneities are actualized and increasingly complex forms appear (galaxies, stars, planets, living creatures).

- **final cause** corresponds to **third quadrant** events. It must be read through both the extrapolated devolution[120] scenario of the mere material substance *and* through the interaction of matter and sentient beings. Matter is attracted towards black holes and might eventually disappear with proton decay. Inside of – and in parallel with – this devolving process, an evolving process develops through a long-term survival of life and intelligence that may take control of the universe. This double process, investigated by physical eschatology, corresponds to the universe's final cause.

- **efficient causality** is closest to the modern notion of cause which, in the historic fresco of the cosmological narrative, tends to be pushed back towards the zero time of the cosmological scale; namely, towards **fourth quadrant** quantum-gravity events, 'when' the primeval 'symmetric' vacuum prevails. It is perhaps the 'place' of the universe's efficient cause 'before' inflation. And yet without confusing the contemporary vacuum of physics with the philosophical void, nor contemporary atoms with antiquity's atomic conceptions, it is interesting to note that Abderites Leucippus and Democritus associated efficient cause with the void[121] in which they placed the primitive movement of atoms.

Thus, as the cosmological sequence of events unfold, structures are made and unmade, particularize and generalize. From broken symmetry to broken symmetry, from phase transition to phase transition, essence passes into existence through gaps and attractions.

The transforming sequence of the universe also refers to alchemical work (in Latin: *Magnum Opus*) which typically involves four steps characterized by the original colours mentioned in Heraclitus: *melanosis, leukosis, xanthosis*, and *iosis*.

- **Nigredo**, Melanosis, 'blackening' or 'cooking' work is the initial state of primordial matter of the *first quadrant*.

- **Albedo**, Leukosis, 'whitening' or 'washing' work is the differentiation process of ego from unconsciousness. It corresponds to the *second quadrant*, which, with matter/light decoupling sees inhomogeneities growth – potential ego consciousness centers – becoming stars fueled from the background of interstellar cloud.

- **Xanthosis**, Citrinitas, 'yellowing' work represents the surmounting of conflicts and contradictions through a sublimation process. There is a drive for separation and objectification which leads to the emergence of consciousness in the *third quadrant* where ego and non-ego confront each other.

- **Rubedo**, Iosis, 'reddening' or 'incandescence' corresponds to Planck era of *fourth quadrant*. This is the final step of the alchemical process where everything comes together in the unity that transcends consciousness.

If a fourth is to be dwelling within any tetrad, what about the (3 + 1) Aristotelian causes and the (3 + 1) steps of the great work? Before coming back to this question at the end of the next chapter, the self-emerging organizational quality of the Three must first be investigated. Intertwined with the quaternary rhythm, cosmological becoming seems to keep pace with the ternary rhythm of the craftsmen-builders of complexity represented by the known interactions of contemporary physics. Here, a force acts, then another retroacts in an inner contradiction that allows for the dynamic transformation of the universe. The ternary rhythm of dialectics comes to mind, proceeding by successive negations: from thesis to antithesis into synthesis.

ENDNOTES

104   Lincoln Barnett, *The Universe and Dr. Einstein, A Clear Explanation of Einstein's Theories*, 2nd ed. (New York, NY: Harper and Row, 1966), 19.

105   von Franz, *Creation Myths*, 254.

106   Edward F. Edinger, *The Bible and the Psyche: Individuation Symbolism in the Old Testament* (Toronto, Canada: Inner City Books, 1986).

107   Paola Zizzi, *The 'Big Wow' Theory*. http://www.quantumbionet.org/eng/index.php?pagina=60

108   Jack Tuszynski, ed., *The Emerging Physics of Consciousness*, (Berlin, Heidelberg and NewYork: Springer, 2006), 16.

109   Giuseppe Vitiello, "The aesthetic experience as a characteristic feature of brain dynamics," *Aisthesis, Vol.8 no. 7* (Florence, Italy: Firenze University Press, 2015), 77. http://dx.doi.org/10.13128/Aisthesis-16207

110   Stuart Hameroff & Roger Penrose, "Consciousness in the Universe: An Updated Review of the 'Orch OR' Theory," *Biophysics of Consciousness: A Foundational Approach*, ed. R. R. Poznanski et al. (Singapore: World Scientific Publishing, 2016) Available at http://www.consciousness.arizona.edu/documents/Hameroff-PenroseUpda tedReviewofOrchOR2016b2237_Ch-14_Revised-2-3.pdf

111   In fact, long before, at an age of 60,000 years, invisible dark matter that does not interact with light condenses into the first dark galaxies.

112   René Guénon, *The Reign of Quantity and the Signs of the Times*, trans. Lord Northbourne (Hillsdale NY: Sophia Perennis, 2001) 40.

113   J.-M. Lévy-Leblond, "Did the big bang begin?" *American Journal of Physics* 58, (1990):156-9. Available at https://www.researchgate.net/profile/Jean-Marc_Levy-Leblond/publications.

114   Roger Penrose, *Fashion, Faith, and Fantasy in the New Physics of the Universe*, (Princeton NJ: Princeton University Press, 2016), 380.

115   Alexandre Koyré, "Perspectives sur l'histoire des sciences," in *Études d'histoire de la pensée scientifique*, (Paris: Gallimard, 1973), 399. [trans. by Nick Jardine, "Koyré's Kepler / Kepler's Koyré," *History of Science* 38 (2000): 363–76, Available from http://adsabs.harvard.edu/full/2000HisSc..38..363J]

116   Lévy-Leblond, ibid.

117   Stephanie L. Hawkins, "William James, Gustav Fechner, and Early Psychophysics," *Front Physiol.* 2011; 2: 68. Published online 2011 Oct 4. doi: 10.3389/fphys.2011.00068

118   Rodney Collins, *The theory of Celestial Influence*, (New York, NY: Samuel Weiser,1971), 156.

119   Eric Chaisson & Steve Mcmillan, *Astronomy Today, 6th ed.* (New York, NY: Pearson Education, 2008),735, Figure 27.1. Available at http://www.as.utexas.edu/astronomy/education/fall09/scalo/secure/301F09.Ch27slides.pdf

120   Evolving and devolving processes will be explained latter in chapter 9

121   André-Jean Voelke, "Vide et non-être chez Leucippe et Démocrite," *Revue de théologie et de philosophie* 122 (1990): 341-52. Doi.org/10.5169/seals-381417.

8. Traces of a Ternary Rhythm in Cosmological Events

Three moments of dialectical processes. Physical interactions as craftsmen and gardeners of the universe. Threefold carving sequences. Reflections of a '4x3' numerical matrix.

8.1. Three Forces and Three Ternary Sequences

Three of the four known interactions of physics (weak nuclear, strong nuclear, gravity, and electromagnetism) are responsible for the construction of structures.

It is the short range of the weak nuclear interaction, and above all its lesser strength compared to the strong nuclear force, which prevents this force from contributing to a particular structure. However, weak nuclear interaction probably played a key role in the first moments of the universe and expresses itself in the transmutation of unstable atoms during radioactive decay. This interaction is responsible for neutron decay into protons and interactions of neutrinos with matter. It is also involved in fusion phenomena in the core of stars as well as in their final evolutionary phases.

Strong interaction holds together charged protons of the same positive charge within the nucleus and is responsible for the stability of nuclei as well as that of protons and neutrons.

The gravitational interaction is the dominant force at the cosmological scale that affects all forms of matter and energy. It is an attractive force whose intensity decreases with the square of distance. Its relative intensity is very weak, but it holds together large structures such as planets, stars, and galaxies, and determines the ultimate fate of the universe.

Electromagnetic interaction (which unifies light with electric and magnetic fields) makes possible: interactions of electrons with nuclei to form atoms, interactions of atoms to form molecules, and interactions of molecules with each other to form solids and liquids. This force is attractive or repulsive, depending on the sign of electric charges. Its intensity – relatively larger than that of gravitational force – decreases, too, with the square of the distance. It transports photons from the sun, from stars, and triggers brain synapses.

A ternary rhythm seems to appear and seems willing to repeat in each quadrant, marked by the development of structures related to the action of forces. A gravity-nuclear-electromagnetic sequence may be detected, at least through the first three quadrants – the fourth may have to wait for a quantum theory of gravity. (Fig. 22)

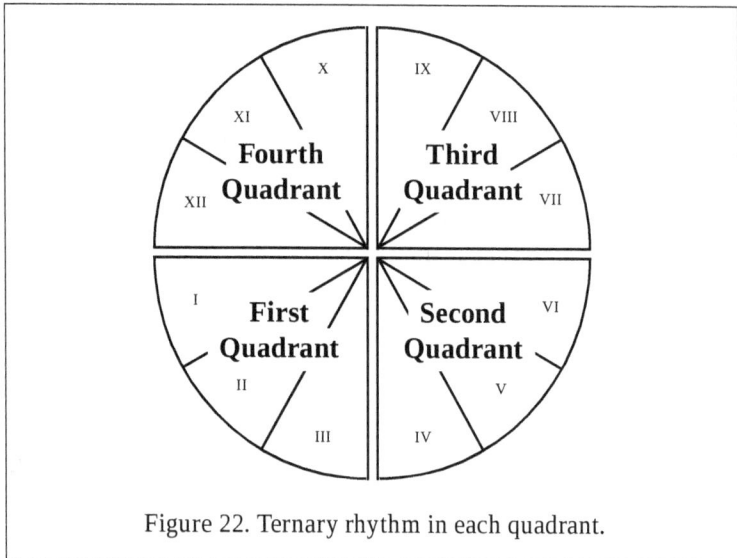

Figure 22. Ternary rhythm in each quadrant.

The gravitational interaction, attractive with extremely low intensity and infinite range, acts locally in IV on matter inhomo-geneities consecutive to light-matter decoupling. It acts globally in X when it emerges victorious in the *Big Crunch* or is defeated by indefinite expansion caused by vacuum energy. It may seem speculative to associate it with cosmic inflation of phase I and to the emergence of reflective consciousness in VII; but the already mentioned Orch OR theory gives account of the two events by involving gravitational interaction through the difference between mass distributions of quantum states. This difference will be explained away but briefly, this means that in phase I, the extreme weakness of gravity compensates for the mass of the 'universe-to-be' to give the duration of inflation; similarly, in phase VII, this same weakness applied to coherent mass displacement in brain tubulins gives a decoherence time in keeping with the best measurable correlate of consciousness, which is the gamma brainwave frequency.

The **strong nuclear force**, which is always attractive, holds the nucleus together by binding protons and neutrons and maintaining the quarks inside protons. Very limited in range, it concentrates the initial energy boost into matter by confining the quarks during the primordial nucleosynthesis of phase II. In V, in stars centers – where thermal pressure opposes gravitational attraction – nuclear reactions 'flare up,' forming carbon and, during supernova explosions, the strong nuclear force creates other elements necessary for life such as those needed to make DNA, the molecule that carries genetic information. In VIII, weak nuclear interaction is responsible for matter disintegration in very different modes and speeds. The absence of a quantum gravity theory does not allow any hypothesis on how these interactions act in phase XI, which in Planck-era conditions are indistinguishable from gravitational and electromagnetic interactions.

The **electromagnetic interaction** occurs between electrically charged particles. It is attractive with opposite-sign charges and repulsive with same-sign charges. In phase III, the electromagnetic interaction binds negatively charged electrons to positively charged nuclei to form hydrogen. In phase VI, electromagnetic interaction form molecules that are at the basis of all chemical and biological phenomena. In phase IX, in the deep, post-material future of the universe, the infinite-in-range electromagnetic interaction connects electrons to positrons to create positronium atoms. In phase XII, the electromagnetic interaction, indistinguishable from the other three, generates electron-positron pairs that take the form of vacuum fluctuations. (Fig. 23)

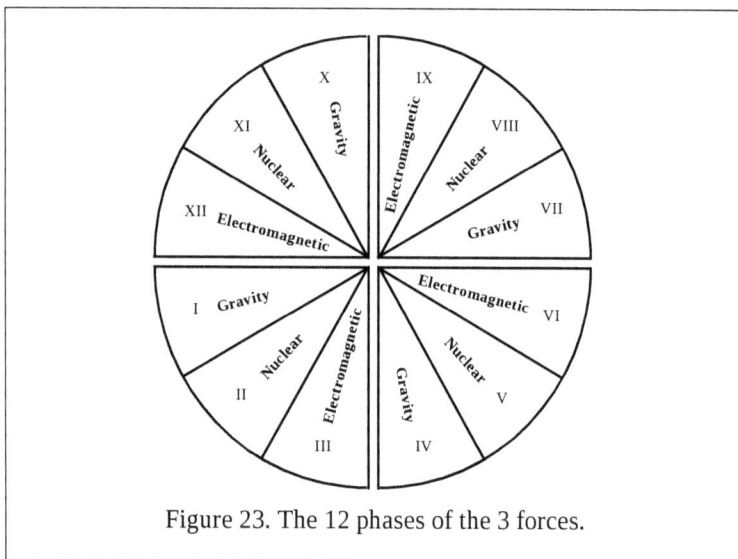

Figure 23. The 12 phases of the 3 forces.

Ternary Sequences.

A ternary rhythm, impressed by physics interactions, is thus articulated within the four quadrants' quaternary rhythm, already defined in the history of the universe. Each quadrant can be divided into three stages that can be considered, at least initially, as the three moments of a dialectical process. Although Jung goes beyond and enriches the Hegelian model of dialectical process – including a much deeper and transmuting level of being – we'll leave it to Hegel, knowing that a *well understood* Hegelian synthesis is very close to the Jungian model.

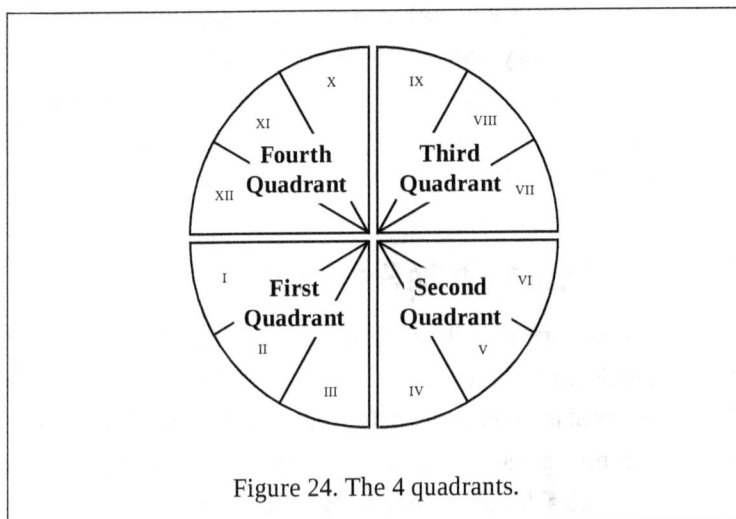

Figure 24. The 4 quadrants.

In the early moments (I, IV, VII, X), the thesis that arises involves gravitational interaction that asserts either directly because of its attraction (IV, X), or indirectly through triggering the *Orch OR* process that connects (I, VII) events to space-time geometries.

In the second moments (II, V, VIII, XI), antithesis confronts as an active negativity, as an opposition act which involves strong and weak nuclear interactions.

In the third moments (III, VI, IX, XII), synthesis solves the confrontation between the gravitational interaction and nuclear interactions through electromagnetic interaction. The latter makes the universal substrate elements to associate, to strike, to weld one to the other, and erase in an assumption of fulfillment.

Thus, throughout the cycle, opposites coexist, interact, and intermingle in an eternal triple-and-four time-intertwined movement, where destructions and constructions interconnect until reaching a new convergence. This new imperfect unity – default or incompleteness – then becomes a new initial thesis that calls for a new confrontation between matter and antimatter, information and entropy, matter and light. Let us now explore the possible reflections of ternary sequential patterns, leaving aside for a later chapter the question of how a return to the whole may be perceived in *each ternary quadrant* as suggested by the mystical precept of 'the One as the fourth.'

8.2. Ternary Sequence in the First Quadrant

Space-time is born in a dazzling burst of energy during a period, called inflation, which starts at time 10^{-35} s and ends at time 10^{-32} s. In all theories of inflation, growth is exponentially rapid. The universe emerges from a quantum vacuum as a phase transition from a meta-stable symmetrical 'false vacuum' to the stable symmetry-broken 'true vacuum.'

At the end of this exponential expansion, where dimensions are multiplied by a factor of 10^{60}, the enormous energy corresponding to vacuum energy is released in every cubic centimeter of space, warming the universe by almost 10^{27} K. All kinds of particles are created – quarks, electrons, and neutrinos – and antiparticles: antiquarks, positrons, and antineutrinos. In summary, we have three steps:

(I) *Affirmation* of a first 'pre-material' luminous moment, during which space-time is created, endowed with a con-siderable energy. Driven by a hypothetical field that could be the Higgs boson, the universe expands with a factor of 10^{60} in an extremely short amount of time (10^{-32} s).

(II) *Confrontation* matter/primordial field. The tempera-ture drop that accompanies expansion slows the initial momentum and condenses the colossal energy of inflation into quarks, leptons, and bosons. The strong nuclear force converts photons into massive particles that *oppose* the energy of inflation.

(III) 'Friction' between matter and light leading, after 380,000 years, to the *synthesis* of the first atoms. Throughout all this phase, photons, which are vectors of information and conveyors of the electromagnetic interaction, remain imprisoned in matter. It is only with the temperature drop that they will be able to freely circulate and communicate.

8.3. Ternary Sequence in the Second Quadrant

The matter / light decoupling ushers the beginning of a stellar age. Inhomogeneities or local matter overdensities (detected today, cooled in the background of the sky) can grow. Several scenarios invoke the structure formations in which cold and warm dark matter intervened.[122] Gravity collects, condenses, and opposes the increase of thermal pressure, which causes overheating that increases speeds and collisions of particles.

Locally, close to the star centers, thermonuclear fusions create all the chemical elements, from helium to iron. During supernova explosions, elements essential for life, such as carbon, are created. Thus, when the universe is 8 billion years old, the solar system forms from cosmic detritus. Photons propagating from nuclear fusions in the core of stars reach the surface after a journey of about one million years and spread into the environment.

On our planet, aggregated 4.6 billion years ago, a local 'friction' between matter and light takes place. Once the surface is cooled, the photon streaming of the sun leads to the development of a dynamic which makes life possible. The electromagnetic force, responsible for chemical reactions, creates bonds between molecules, some of which acquire characteristic properties of life. In summary, we get back to three stages, closer here, to Schelling's primitive nature with his 'primal seeds' hypothesis and his distinction within the absolute between ground (Grund) and existence (Existenz). His metaphors are very appropriate for the second cosmological quadrant: gravity versus light, darkness versus light[123], the 'luminous principle' being for him, the 'second potency' of Nature:

> (IV) *Affirmation* of a first moment when the attractive force of gravity leads to growth of local inhomogeneities or local overdensities of matter, which are potential homes for galaxies to come and centers for any object: star, comet, animal, human, etc.

> (V) Thermal pressure *opposes* to gravitational attraction towards the cores of stars. These high-density areas will warm more and more, causing conflict between the nuclear and gravitational forces. Thermonuclear fusions make billions and billions of stars explode or implode, which gradually enrich the environment with carbon atoms and heavy atoms.

> (VI) The local 'friction' between matter and light leads to biosphere growth. Elements synthesized in stars and dispersed into the interstellar medium are *combined* on Earth. The union of matter and light leads to the development of a dynamic which makes life possible. Molecular interactions having an essentially electromagnetic nature develop forms which become better and better adapted to the environment, ensuring the communication of information around the membranes of the cells of biological beings.

8.4. Ternary Sequence in the Third Quadrant

No scientific model claims to explain or predict everything, and all are speculative to varying degrees, whether they be launched over the very distant past or toward the deep future of the universe. With the third quadrant, we enter the field of physics extrapolations: that which is beyond what is firmly established and that which may not respect all the normative prescriptions set out by Karl Popper.[124]

In the path first laid out by Jung and Pauli, the degree of speculation of 'conceptual experiments' is of little importance insofar as the laws and methodology of modern physics are respected. As the third quadrant starts with the 'emergence of reflective consciousness' event, this latter phenomenon is endowed with a fundamental role in the universe as a whole, which is a minority view in contemporary cosmology. Arriving in the stellar era of a universe 13 billion years old, *Homo sapiens* is considered as an epiphenomenon by most cosmologists. This appearance is seen as a trivial event that will disappear as quickly as it appeared, most likely without leaving trace.

Nevertheless, the human brain is the last biological frontier, at the summit of universal complexity. We will follow here the optimistic thought of those who not only see consciousness as maintaining itself in the stellar era of the universe, where matter undergoes an overall entropic evolution, but also in the period when it could play a fundamental role: namely, during the degenerate era when stars turn into white dwarfs, brown dwarfs, and black holes. And even in the period of proton decay and beyond, into the dark era that will continue indefinitely with unique 'material' structures represented only by electrons and positrons.

For a diachronic reading of the third quadrant, we will not use here already mentioned quantum models of consciousness, but instead a phenomenological approach. We will rely on empirical research of paleoanthropologists, which exhibit a strong correlation between thinking that guides action and technical gesture. Technique appears as a constitutive element of man's condition and evolution.

For André Leroi-Gourhan, the evolution of the body, the transformation of the environment by technique, and the transfiguration of the environment by the symbol, have not only kept pace, but have also been co-generated. He showed how the 'sparks of consciousness' that came to light from purely organic origins have arisen 'from the biological to the technological and socio-cultural' in an exponential evolutionary acceleration. This transition appears in the graph of the change of power as a function of the successive forms of engine power available to man. (Chart 25)

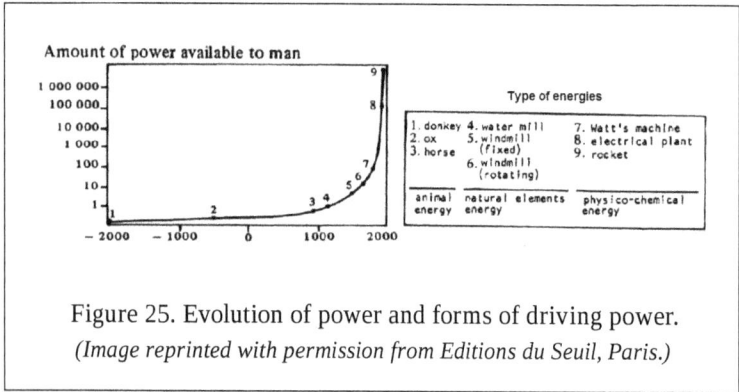

Figure 25. Evolution of power and forms of driving power.
(Image reprinted with permission from Editions du Seuil, Paris.)

The composite chart below shows the temporal evolution due to both the increased volume of the brain and the progress of prehistoric 'industry.' [125] (Chart 26)

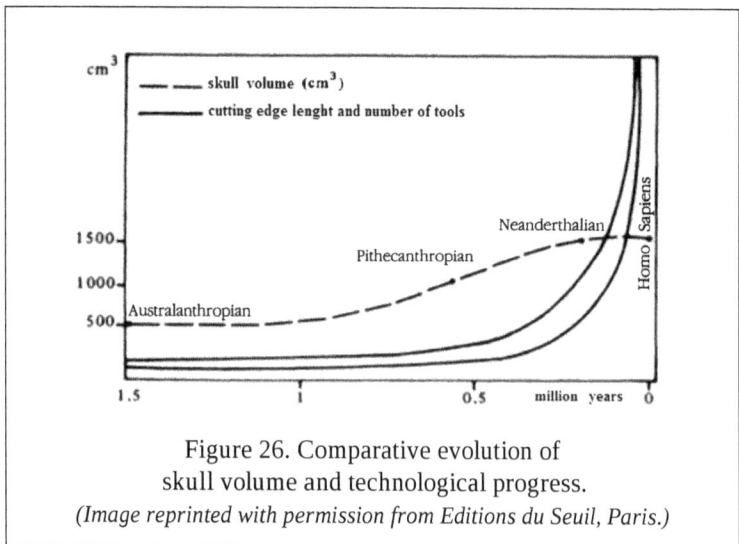

Figure 26. Comparative evolution of
skull volume and technological progress.
(Image reprinted with permission from Editions du Seuil, Paris.)

First extremely flat, technological curves gradually accelerate before drilling the brain curve just when the latter comes to cap. François Meyer writes:

> It is as if the bio-anatomical variable, having reached the limit of its evolutionary possibilities and unable to continue the pace of acceleration, saw itself relayed by another variable, endowed with less inertia and able to take over. Leroi-Gourhan puts his finger on the issue of the major relay by which evolution jumps from bio-anatomical orbit to technological orbit [...][126]

And he adds, for the same reason, that

> [a]t the other end of history, we see now an eye-catching relay. The conquest of energy, which dominated history, today sees its millennium progression reach its limit. The energetic variable is relayed by another variable, endowed with remarkable rate of growth, the informational variable. It is the latter that mobilizes economic investment and technical invention. It is on the latter that evolution today relies.[127]

This biologizing conception of technology, which is the tendency that "pushes handheld flint to acquire a handle, the bale dragged on two poles to carry wheels,"[128] converge with the concerns of transhumanists, who see biotechnological development of humanity as a vital necessity.

Physicist Freeman Dyson argues that: "Life resides in organization, not in substance."[129] Assuming the basis of consciousness is not associated with a particular type of matter such as carbon, but instead with the complex structure of the substrate, he showed that life could persist in the cold and dark conditions of an open universe in indefinite expansion. After 100 billion years, life and intelligence would thus be able to live on long after the disappearance of *Homo sapiens.* Indeed, consciousness could be incorporated into a cloud of particles bearing positive and negative charges, both organizing and communicating by means of electromagnetic interactions.

Frank Tipler has taken up the Dyson arguments and extended them to the scenario of a closed universe. He showed that all matter of the universe could be transformed into a huge information processing apparatus. When the universe would collapse on itself (Big Crunch), the computational capacity of the universe will have become infinite. This will be the state of the Omega Point, when intelligent life will have the attributes of 'omnipresence, omniscience, and omnipotence.' However, scenarios corresponding to current dark energy measurements show that the universe's expansion may accelerate to such an extent that the electromagnetic force, which holds things together, would have no effect and any object would be 'torn apart': *Big Rip.* In these circumstances, Tipler envisions the intervention of consciousness on the overall behavior of the universe, which would force it to contract again.

In summary, like before, we find three phases.

(VII) *Assertion* of reflective consciousness, which is a vibrant part of cosmogenesis, emerging in an exponential growth phase. Gravity plays a role in this emergence since, according to Leroi-Gourhan, the upright posture of bipedalism released man's hands and allowed him to technically invent and access culture. Retained by gravity, the Earth's atmosphere is a further decisive factor for the emergence of the *noosphere* or biosphere 'thinking stratum.'[130]

(VIII) The disorganization of material substrates and proton disintegration into lighter particles (electrons, neutrinos, photons), *confronts* conscious beings and challenges them to regenerate and proceed through continual metamorphosis.

(IX) Following a total transmutation of the universe, evolutionary *synthesis* towards a unity of consciousness occurs. Photons, information carriers, and conveyors of the electromagnetic interaction connect electrons and positrons on which conscious beings are coded. Electromagnetic interaction extends all communication networks and tends to saturate the universe into information while eliminating horizons.

8.5. Ternary Sequence in the Fourth Quadrant

In the first three quadrants, the ternary dialectical processes were more or less based on the chronological scale of physical time. The very sequence of the first three quadrants seemed to reflect the symbolism of the first three numbers: the outpouring of undifferentiated energy in the first quadrant, separation between light and matter in the second, and emergence of an 'observing consciousness' which, as von Franz observes, inserts a symmetrical axis into the two-rhythm, or else 'counts' the latter's temporal and spatial succession.

Now, the fourth quadrant is associated with Planck's conditions of extreme temperature and density. There is no clear definition for a linear physical time nor even for space-time. Advances have been made by physicists in considering possible proto-times and proto-spaces associated with quantum fluctuations at the 'zero-point energy' of the vacuum that may let catch a glimpse of 'moments' of a dialectic. In a following chapter, building on an overview of an overall structural symmetry, a possible three-fold dialectical process underlying the 'Planck quadrant' could, to some extent, be deduced by analogy from the opposite 'IV-V-VI' events that benefit from a better knowledge since they are related to well controlled physics and chemistry in the laboratory. (Fig. 27)

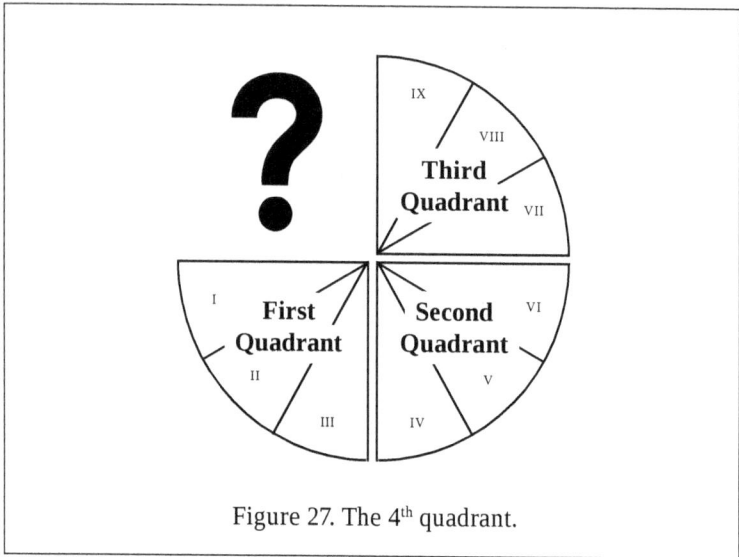

Figure 27. The 4th quadrant.

Nevertheless, let us try to imagine the mechanism behind the work of the forces of physics in the fourth quadrant. The gardeners of the universe who meticulously carved a ternary process in the first three quadrants have become indistinguishable under Planck-era conditions. Although thought to be melted into the three other interactions, always-attractive gravity *acts* on the 'quantum foam' of empty space.

This could eventually represent the first phase (X) of a ternary process à la Schelling, protecting the seed of a new pledge of a universe in the folds of its spatial curvatures. This gravitational *pull* quite matches the hypothetical 'previous' final point-like singularity, or bouncing universe of the 'late' third quadrant. Under extreme Planck-era conditions, the universe would probably consist of 'global information' that could be handled in-depth by recent theoretical approaches explaining gravity as an entropic force.[131] This supposes a connection between gravitation and information, metaphorically illustrated by the human tendency for people to gather in cities in order to benefit from still more information.

In the second phase (XI), the *repulsive* reaction of the 'always-already-there' quantum vacuum pressure could be the generating mechanism behind the superpositions or separations within the wave function of the universe – i.e., an actual infinite number of possible curvatures of any form associated with an infinite number of potential universes.

The third *synthesis* phase (XII) could see the infinite number of quantum superpositions 'erased' in a few Planck times and be reduced to a *finite* quantity of entangled universes.

8.6. The Unfinished Game of the Untotalizable '4 x 3' Structure

If we consider again the three forces that are responsible for the construction of structures, not as actors of the four ternary dialectical processes but in terms of their 'aspiration,' we can see that the universe *had* to follow a circular causal path. For, had the forces been guided by the search for the state of lowest possible stability along a linear model of causal path, then, out of the 'initial' energy boost, gravitational force would have formed black holes, strong nuclear force would have formed iron nuclei and electromagnetic force would have formed rare gases. Obviously, all by-products of stars, nuclei, atoms and molecules could not have formed, and reflexive consciousness could not have emerged. Instead, vacuum fluctuations and all potential inhomogeneities would have been transformed into gravity's ultimate-victory black-holes while the strong nuclear force would have yielded a universe of iron. But, a certain rate of expansion of the universe (related to the Hubble constant) seems to be constantly 'instilled' by the fluctuating space-time. These quanta of gravity (equivalent to the yet unobserved gravitons) are not immersed in space, they *are* space, maybe co-created by cosmic gravitational events that act as a circular causal way on feed-back loops through the whole of the four quadrants. The undifferentiated fourth quadrant points toward the mystery of gravity as the force which, not only builds stars and galaxies at local level but also, in a yet unknown competition mechanism with dark energy, governs the rate of expansion at global level. This rate sets up – at the onset of each quadrant – phases of non-equilibrium states in such a way that the works of the forces remain *unfinished*.

Wherever they dwell, the laws of physics engage a dialectical play with ubiquitous chance and this is another conjunction of opposites. Between order and disorder, a game is played in the universe like a kind of reinventing the game of Aion (cyclic time or eternity?) evoked by Heraclitus through his famous fragment: *Aion is a child playing draughts; the kingship is the child's.*

8.7. The Fourth of the Aristotelian Causes and Alchemical Process Revisited

In a letter to Pauli, Jung recalls "The transition from 3 to 4 is often an awkward matter, in that according to historical symbolism either the three deny the fourth or the fourth neutralizes the three into a whole (as in the axiom of Maria, for example)."[132] As the 'game' of the universe interpreted from the cosmological scientific model allows for some timid reflections of an ordered four-quadrant matrix to shine through, the 'fourth' or undifferentiated and underdeveloped function seems to be dwelling in the current difficulty to overcome quantum gravity events of the fourth quadrant.

But just as the dominant function – in Jung's theory of psychological types – may fluctuate throughout the course of a lifetime, the inferior 'fourth' function may change within the history of ideas and concepts. In *Psychology and Alchemy*, Jung mentions that

> [...] about the fifteenth or sixteenth century, the colors were reduced to three, and the *xanthosis*, otherwise called the *citrinitas*, gradually fell into disuse or was but seldom mentioned. [...] Whereas the original tetrameria corre-

sponded exactly to the quaternity of elements, it was now frequently stressed that although there were four elements (earth, water, fire, and air) and four qualities (hot, cold, dry, and moist), there were only three colours: black, white, and red.[133]

For James Hillman, colors, no less than numbers, must be considered archetypal powers. Taking up the question of the omitted *xanthosis* (*citrinitas* 'yellowing') within the context of alchemical relations, he considers that yellow appears as a specific transitional quality in a temporal process from interiority as experienced in the white stage [second quadrant], to an expanding awareness through turning our attention outward to the world [third quadrant].

> Having absorbed and unified all hues into the one white, the mirror of silvered subjectivity expands to reflect all things at the expense of differentiation of itself. It takes something outside subjectivity to see into oneself, [...] anything outside subjective reflection, any moment that intrudes upon white consciousness's love of its own lunar illumination, which is precisely where its blindness lies.[134]

Clearly for Hillman, the yellowing is more than a "spoiling of the white." But even when Gerhard Dorn mentions the form or intellect as "made clear by the saffron color,"[135] to Hillman's eyes, this pupil of Paracelsus does not describe adequately the nature of the yellowed intellect.

> It is not the continued expansion of white consciousness, an increase of reflective capacity. It is more than aware, more than enlightened. It must be hot and male, beginning with the light yellow or dirty yellow-brown of Mars and even putrid, if it is to spring the mirrored prison of reflection. Since sulfur with all its corruption, intensity,

and feculence has been instrumental in its change, this mind 'burns' with its own bitter bile, yellow choler.[136]

As James Hillman restores a crucial significance to the omitted yellow, he simultaneously brings out the quality of the 'fourth' to third quadrant events of the universe where modern cosmological science has difficulty accounting for the jump of "white ego consciousness" – in second quadrant – over the threshold of reflective (and unhappy) consciousness of third quadrant. Most of cosmological models do not account for the reflective capacities of the human brain and even less the possibility of sentient beings in the future controlling the development of the universe.

This uncertainty of the inferior function is also shown by the dismissal of final causation (found in third quadrant) that occurred at the same contemporary epoch of the missing alchemical stage. The restricted notion of causality adopted with the rise of modern science led to the dependence on a deterministic chain of past events that remains essentially unchallenged in contemporary science. This way of describing the natural world has however been called into question. As already shown by Pauli, observations in quantum physics force us to abandon Cartesian dualism. Fludd's quaternary approach supports the role of the *observer* in the observed phenomenon. Even if the idea that consciousness would play a major role in the universe is not fully accepted within the scientific community, John Wheeler has nevertheless suggested that everyone of us are acting through quantum-mechanical acts of observation, a process he calls 'genesis by observership': "We are participators, at the microscopic level, in making that past, as well as the present and the future."[137]

However, the *observer* in the vigorously criticized theory of Frank Tipler does not lead to a comparable reciprocal self-engendering. The omega point is 'the ultimate' observer – *outside* the universe – responsible for bringing the entire universe into existence. On the other hand, in Roger Penrose's Objective Reduction theory, there is *no more observer* since quantum wave function self-collapses, a verb which, in the wake of Jung and Pauli ideas, could be written with a capital 'S.' The observer disappears behind protoconscious events (collapses of the Self?) which have a universal character, being encoded in space-time geometry and taking place at each level of the universe.

Thus, as more and more scientists dare to model the future evolution of the universe, expanding the narrowly construed causal system to include the possibility of sentient beings, Jung's aims for alchemy, "the rescue of the human soul and the salvation of the cosmos,"[138] may come true.

ENDNOTES

122   Houjun Mo et al., *Galaxy Formation and Evolution*, (Cambridge, UK: Cambridge University Press, 2010), Available online at https://www.astro.umd.edu/~richard/ASTRO620/MBW_Book_Galaxy.pdf

123   Alistair Welchman, "Schelling's Moral Argument for a Metaphysics of Contingency," *Nature and Realism in Schelling's Philosophy*, ed. Emilio Carlo Corriero and Andrea Dezi (Turin, Italy: Accademia University Press, 2014), 359.

124   Popper's main criterion to distinguish scientific knowledge from other source of knowledge is *falsifiability*. A theory is scientific only if it can be refuted or shown false by observations.

125   François Meyer, "Temps, devenir, évolution [Time, Becoming, Evolution]," *Communications* 41 (1985): 111-22. [translated by the author] DOI : 10.3406/comm.1985.1611

126   Ibid., 116.

127   Ibid.

128   André Leroi-Gourhan, *L'homme et la matière* (Paris: Albin Michel, 2012). [translated by the author]

129   Freeman Dyson, *The Origins of Life* (Cambridge UK: Cambridge University Press, 1985).

130   Irina Trubetskova, "Vladimir Ivanovich Vernadsky and his Revolutionary Theory of the Biosphere and the Noosphere," *INCO 796: Cosmology and Our View of the World* (2004) Available at http://www-ssg.sr.unh.edu/preceptorial/index.html

131   Works such as those of Erik Verlinde and Thanu Padmanabhan, suggest that gravity is not a fundamental force, but could emerge in the way of the gas pressure from collective behavior of molecular motions. This approach could also lead to a better understanding of how matter curves spacetime.

132   Jung to Pauli, October 24, 1953, n° 64J, p. 128.

133 Jung, *Psychology and Alchemy,* § 333.

134 James Hillman, *Alchemical Psychology: The Uniform Edition of the Writings of James Hillman, vol. 5.* (Putnam, CT: Spring Publications, 2010).

135 Jung, ibid., § 366.

136 James Hillman, ibid., who quotes R. Klibansky et al., *Saturn and Melancholy: Studies in the History of Natural Philosophy, Religion, and Art* (New York, NY: Basic Books, 1964), 53.

137 John A. Wheeler, "This Participatory Universe," in *A Passion to Know: Twenty Profiles in Science,* ed. Allen L. Hammond (New York, NY: Scribners, 1984), 185.

138 Jung, *C.G. Jung Speaking: Interviews and Encounters,* ed. W. Mcguire and R.F.C. Hull (Princeton, NJ: Princeton University Press, 1977), 228.

9. The Archetypal Images of the Zodiac as Projection of Numbers-Archetypes

Twelve zodiac signs as projections and symbolic vestures of the '4x3' matrix of numbers-archetypes. Cyclic interplay of the Night and Day forces. Involution, Devolution, and Evolution processes.

9.1. Zodiac Signs Expand Number Symbolism

In order to fully apply the language of numbers, we will now re-use the 4 × 3 matrix of numbers-archetypes while infusing it with the substance of an isomorphic symbolic structure: that of the zodiac. The zodiac is a major symbolic figure with a great metaphorical value encountered in many cultures. The 'Wheel of Life' was represented in Greece by the Oceanos river surrounding Earth, or by Chronos, the god of time, represented by a snake encircling the universe and wearing the zodiac wheel on its back.

It is not so much in the heavens as in the numbers. As the number-archetype, it originates in a space existing by itself, an intermediary between the intelligence of pure spirit and matter. It is based on the subdivisions of the circle by the first integers. As the number-archetype, it is a field of relation, and as such, could help by 'reflection' a science entangled in myriad theories aimed to understand dark matter and energy and other black hole horizons.

As projections of number-archetypes onto constellations or illustrated by seasons, the 12 zodiac signs are archetypal images symbolizing the journey of the soul and the psychological development of people and things in constant transformation. Like alchemy, astrology is a very rich language, used in the first sketches of psychological typologies. It represents a model of thought and logic that perfectly accounts for the structure of the human psyche as a whole.

In any archetypal image or symbol, cultural elements interfere that particularize the manifestation of numbers-archetypes. In this way, while the zodiac has been following mankind since the earliest of times, it reflects through legends or myths such as the Gilgamesh epic, Odysseus' sailing home or Hercules' twelve labors. Around the fourth century B.C., the fixed zodiac of constellations was abandoned in the West, which preferred the mobile 'tropical' zodiac of seasons, whose Aries zero degree (beginning spring) then coincided with zero degrees of the Aries constellation.

But, because of the precession of the equinoxes, the tropical zodiac – derived from the geometric division of the annual orbit of the sun in 12 sections of 30 degrees – has slowly shifted with respect to the fixed zodiac. This is what today makes the sign of Aries correspond to the Pisces constellation, and so on for the following signs. This shift has become a constantly repeated argument by astrology opponents or supporters, ignorant of its symbolic nature claiming the status of a 'natural science' or calling for a return to the 'zodiac of the stars.'

From the beginning of modern science, this confusion between different planes has fed many alleged anti-astrological arguments. For example, that of the seasons cycle, that is reversed in the southern hemisphere, which would therefore abrogate the meanings of signs. Or still even the concern for the physical inequality of the 12 constellations, or the possible existence – and therefore possible physical effect – of a 13th constellation. The result of a scientific contamination of astrology led to a flattening of thought, consider- ations totally alien to the spirit of Ancients who knew that they were dealing with what is today recognized as a figure of mandala: namely, a symbolic representation of wholeness, based on numbers.

It is unfortunate that, like alchemy or numerology, often reduced to pre-chemistry or pre-mathematics, astrology was often limited to mere pre-astronomy. A proper investigation would allow us to realize that these three disciplines represent genuine philosophies, postulating the unity of matter and spirit, going beyond the traditional divide between reason and intuition. And, among the three disciplines, astrology is the most exposed to the wild syncretism mixing spirituality and science.

Indeed, we know today that the ability to turn lead into gold is real and that this process, although scientifically proven, is not to be confused with the spiritual quintessence of alchemy.[139] It is part of nuclear physics and is not used because of the much higher cost than the value of the collected small amount of precious metal. Similarly, the Pythagorean tradition of numbers and arithmetic science are today deployed in separate fields of knowledge.

By contrast, the emergence of astronomy and the rapid progress of celestial mechanics very quickly contaminated astrological symbolic art, as evidenced by astral determinism treaties of the last court-astrologer, Morin de Villefranche, before astrology had been banned as an academic discipline in 1666.[140]

In parallel with this contamination and with the advance of classical physics, the suspicion surrounding astrology with regard to purely scientific work continues to this day. Galileo, Kepler's contemporary, did not admit the scientific influence of the moon phases on tides, which is now scientifically recognized. These prejudices continue, even today, against perfectly scientific work which studies cause-and-effect links of the cosmos on plants, animals, and humans.

Pauli was very aware of the distinction between a symbolic language of astrology and that of physics. He showed Jung his reluctance to accept astrology when the latter used statistics to try to verify his theory of synchronicity through traditional astrological configurations. However, in his book on synchronicity, published in parallel with Pauli's study on Kepler, Jung clearly recognized that he had strayed into a level of reality where scientific tools are inadequate. In fact, he could not convince himself solely of the psychological aspects of astrology, and was inclined as well to consider causal actions, which, in turn, led to many misunderstandings.

This is evidenced by what he stated at the 1950 Eranos session, published one year later:

> As Professor Knoll has demonstrated at this meeting, the solar proton radiation is influenced to such a degree by planetary conjunctions, oppositions, and quartile aspects that the appearance of magnetic storms can be predicted with a fair amount of probability. [...] So, it is probably a question here of a causal relationship, i.e., of a natural law that excludes synchronicity or restricts it ...[141]

But over time, his thought evolved as evidenced by a letter to Hans Benders in 1958, which shows his difficulty "to explain the astrological phenomenon. I am not in the least disposed to an either-or explanation. I always say that with a psychological explanation there is only the alternative: either *and* or! This seems to me to be the case with astrology too."[142]

Like Pauli, Jung was a research scientist. As with UFOs, it seemed normal to him to question the existence of physical links between the cosmos and ourselves which exist beyond a shadow of a doubt. Indeed, his concerns remain largely anecdotal. Confusion in this respect essentially comes from the fact that, at that time, fields of knowledge such as Chronobiology or Chronoastrobiology did not yet exist, whereas today, they are fully fledged scientific disciplines.[143] But confusion on this matter persists today, despite better understanding of these phenomena.

9.2. Jung's Equation of the Self

Jung has largely drawn on traditional alchemical images to illustrate archetypes or complexes, such as the Anima, Animus, Shadow, etc. He felt he had to write about the Self and this had obsessed his mind for several years. Going through the ordeal of a disease in 1944, "I realized it was my duty to communicate these thoughts, yet I doubted whether I was allowed to give expression to them. During my illness, I received confirmation and I now knew that everything had meaning and that everything was perfect."[144] In 1947, he wrote to Father Victor White[145] that, suffering from serious insomnia, he began writing blindly, without knowing where that would lead him:

> Only after completing twenty-five pages in folio that I began to understand that my secret goal, was Christ – not man but the divine being. It came to me as a shock, as I felt utterly unequal to such a task. [...] Continuing to write, I have reached the archetype of the man-god and the correlative phenomenon of synchronicity of this archetype. Thus, I came to the topic of *ichthys* and the aeon which then opened (transition from de $0°\Upsilon$ to $30°\mathcal{H}$).[146]

Christ, this figure of the divine for the West, Jung would illustrate through the zodiac of astrology, here clearly understood in its archetypal aspect. In his book *Aion*, published in 1951, he calls history as a witness by analyzing the correspondence between the *era* or *eon* of Pisces (related to the precession of equinoxes) and the bi-millennial development of the Christian religion. The archetype of the man-god stands in the center of the zodiacal circle; which is to say, the Self that projects into human history the succession of archetypal images becoming predominant one after the other. These are the 12 astrological signs of the precessional cycle, sometimes called the Great Year or Platonic Year.

In the last chapter, entitled "The Structure and Dynamics of the Self," Jung cites the zodiac as a symbol of totality and its remarkable structure (4x3) as a solution to the dilemma of three plus one. However, Jung goes well beyond this archetypal structure, engaging in a fascinating interpretation of historical images of the Self, and comes finally to a diagram shown in Fig. 28. Clearly, the focus is given to number four and to the geometric figure of the square, which symbolizes both the stability and permanence of the Self. But the triad also appears through a dynamic process of restoration and rejuvenation that Jung describes using alchemical symbolism:

> The formula reproduces exactly the essential features of the symbolic process of transformation. It shows the rotation of the mandala, the antithetical play of complementary (or compensatory) processes, then the *apocatastasis*, i.e., the restoration of an original state of wholeness, which the alchemists expressed through the symbol of the ouroboros, and finally the formula repeats the ancient alchemical tetrameria, which is implicit in the fourfold structure of unity: ⟨⟩. What the formula can only hint at, however, is the higher plane that is reached through the process of transformation and integration. The 'sublimation,' progress, or qualitative change consists in an unfolding of totality into four parts four times, which means nothing less than its becoming conscious. When psychic contents are split up into four aspects, it means that they have been subjected to discrimination by the four orienting functions of consciousness. Only the production of these four aspects makes a total description possible. The process depicted by our formula changes the originally unconscious totality into a conscious one. The Anthropos A descends from above through his Shadow B into Physis C (=serpent), and, through

a kind of crystallization process D (=lapis) that reduces chaos to order, rises again to the original state, which in the meantime has been transformed from an unconscious into a conscious one. Consciousness and understanding arise from discrimination – that is, through analysis (dissolution) followed by synthesis, as stated in symbolical terms by the alchemical dictum: '*Solve et coagula*' (dissolve and coagulate). The correspondence is represented by the identity of the letters, a, a_1, a_2, a_3, and so on. In other words, we are dealing all the time with the same factor, which in the formula merely changes its place, whereas psychologically its name and quality change too. At the same time, it becomes clear that the change of place is always an enantiodromian change of situation, corresponding to the complimentary or compensatory changes in the psyche as a whole. It was in this way that the changing of the hexagrams in the *I Ching* was understood by the classical Chinese commentators.[147]

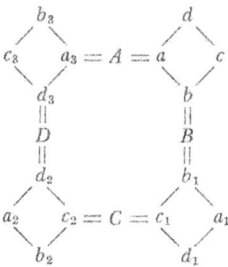

Figure 28. Jung's symbol of the Self.
(Image reprinted with permission from PUP and Taylor & Francis.)

Regarding contemporary cosmology, we will use the symbolic structure of the zodiac, partly used by Jung in his book *Aion*. Building on the tradition of the Platonic Year, Jung analyzes the projection of the Self onto the religious and cultural history of the West. Wherever he lives, man has always had intuitions of cosmic cycles. Hindu tradition, for example, relies on Kalpas, a Sanskrit word meaning "arrangement in time," whose great quaternary universal year lasts 4.32 billion years. Hindu symbols appear in this transcendental history, related to the singularity of the Indian subcontinent religious spirit, symbols that illustrate in their own way the circumambulation of the archetypes around the Self or Brahman. In any case, this is not a particular arrangement (Kalpa) or a particular numeric structure such as the Zodiac that compels history to be what it has been. Rather it is history, created by man, reconstructed by the historical science – or interpreted from a scientific model – which *appears through a symbolic signifier* like the zodiac, taking on specific forms according to the cultures where it arises throughout time. In *Aion*, the historical period was limited to the Christian bimillenary aeon; however, the cosmic narrative aims to interpret all past and future events in the universe. In both parallels, history appears as traces of psychophysical energy that reflect a dynamic expression and regeneration of the Self.

9.3. Illustration of Cyclical Process

The transformation of nature or of a plant during the four seasons of the year illustrates the cyclical process of the zodiac. In *Astrological Signs: The Pulse of Life*, Dane Rudhyar brings the zodiac closer to the figure of oriental Tao.[148] Throughout the annual cycle of solar journey, two forces interpenetrate and alternate in intensity, delimiting four quadrants. The 'Day-Force,' personalizing or Yang energy (white in Fig. 29) begins once more to grow during the winter solstice after a 'stopping-point,' which is the etymological meaning of the word 'solstice.'

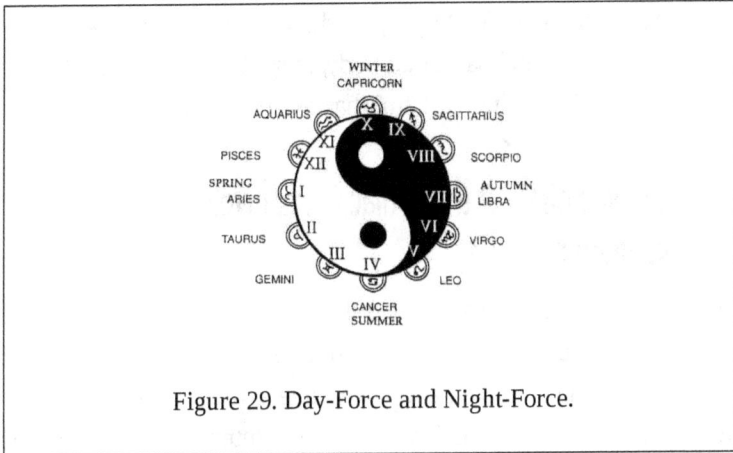

Figure 29. Day-Force and Night-Force.

The 'Night-force,' collectivizing or Yin energy (black in Fig. 29) begins to grow again in the summer solstice, which is the second 'stopping-point' of the solar march. At the spring and fall equinoxes, day and night are of equal length, and the two forces are in a state of unstable equilibrium.

During the day, we only see the Sun; the Day-Force is thus a personalizing energy. At night, we see the company of stars and the Night-Force is thus an in-gathering collectivizing energy, which begins to grow again during the summer solstice in Cancer, the sign of home. It grows through Leo, Virgo, and blossoms personality in society (Libra) by pushing the individual to seek an ever-deeper identification with ever larger communities.

During the winter solstice (Capricorn), the Night-Force triumphs. This is the vast collective body of the state that dominates even its leaders. To illustrate Capricorn, Rudhyar also cites the figure of the eastern yogi who, although begging and living alone, participates in his own way in society. To illustrate the awakening of the personalizing energy of Day-Force – which begins to grow again at Christmas, but becomes clearly visible only in Aries, a germination symbol – Rudhyar uses the figure of the Christ, born in the Roman Empire, a symbol of spiritual incarnation. The Day-Force drives spiritual ideas or entities to take a concrete and particular body, to settle in the center of a personality with the feeling of 'I am' that culminates in Cancer.

9.4. The Two Spirals of the Zodiac Circle: Involution, Devolution, and Evolution processes

We started with the idea of cosmological time being a linear time (past-present-future) allowing us to date events of the history of the universe. For convenience, we have arranged this time on a four-quadrant circle, which allows for the integration of a possible cyclic process. The latter is symbolized by the zodiac, where beginning and end merge while never repeating identically.

We never bathe in the same water, even if it's in the same river. Moreover, neither the history of the universe, nor individual or collective man's histories return eternally in the same way. Hence the idea of including the cycle into the spiral symbol. Physics reflects a 'vertical' time expressed by the second law of thermodynamics, which posits that physical energy is inexorably deteriorating during its cosmic becoming while increasing entropy.

Generation and corruption coexist in the natural environment, hence the double spiral symbol that synthesizes both cyclical time of a circle and linear time of a straight line. Once the first circle is completed, there is a renewal on another level with a growth at each cycle, represented by a double spiral. The zodiac symbol is thus 'thickened' by completing the two-dimensional form with a perpendicular axis around which two spirals are rotating. (Fig. 30)

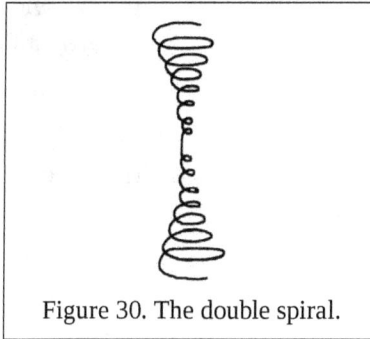

Figure 30. The double spiral.

The two spirals, one going downwards and other upwards, reflect "a flowing in two opposite directions which meet and permeate each other in what is simply and solely *One*."[149] All can be integrated into a sphere, the zodiac wheel being thus placed in an equatorial plane (represented in Fig. 41). This spherical structure is similar to Raymond Abellio's absolute structure[150] that uses a Christic vocabulary to designate the vertical movement: the embodiment for downflow and the assumption for upflow.

The analogy with a plant transformation through seasons illustrates the two vertical axis flows. From the first half of the annual cycle (in correspondence with the autumnal Libra), when the plant reaches its 'flowering,' new 'seeds' form that give rise to the new cycle in the next 'spring,' while 'leaves' (inevitable by-product) decompose to provide the raw material of the new cycle.

Life – in our example, that of a plant – thus evolves in an opposite direction to matter evolution, towards greater complexity and order. Matter evolves towards less order and greater simplicity in the direction of increasing entropy. In many traditions, these 'vertical' processes carry the names involution, evolution, and devolution. The first half of the cycle, referred to as involution, corresponds to the incorporation into matter of an impulse or seed idea; a spiritual potentiality that will express itself through the multiple developing forms and structures. The second half of the cycle, referred to as evolution, corresponds to the development of consciousness. In parallel with the evolutionary phase, the devolutionary phase is the converse of involution and corresponds to the degradation of the material substratum, which is a degenerative process that exists in all creatures.

Spirals or twin-serpents symbolize those principles or simultaneous flows of cosmic energy that are present in all things. The Tao expresses them through the sinuous forms of Yin and Yang that symbolize the perfect figure without beginning or end. They are present in many traditions, especially among Neoplatonists, who revived Platonism while absorbing ancient currents that existed in the third century (Stoicism, Aristotelianism, Skepticism), until the Renaissance.

In Neoplatonism, everything emanates from the absolutely ineffable and transcendent One that Plotinus, the first Neoplatonist (205 - 270 A.D.), identifies with the Platonic Idea of the Good. The One is the condition of possibility from which all beings derive. It does not participate in Being, for it is "beyond Being" or "barred One." (Fig. 31)

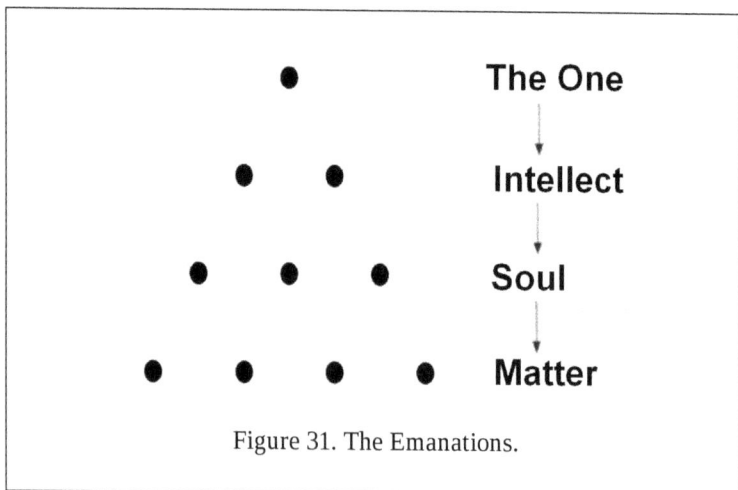

Figure 31. The Emanations.

Then comes the Intellect (Nous), the One that 'is,' to which corresponds the dyad of the Being and of the Intelligible Realm. Proceeding from the Intellect, then comes the Soul (Psyche), the intermediary between the intelligible and sensible worlds, of which it is the organizing instance. Animated by a circular and centrifugal movement, Soul, by way of its mediatory nature, is composed of three parts or aspects: the higher part contemplates the intelligible, the lower part is in contact with the sensible, and the third, which acts as an intermediary, comprises sensation and reason.

Below the three hypostases is a 'fourth' amorphous reality, Matter, in which our individual soul is trapped. Matter is made up of the four elements. The elemental fire receives some sort of shape and, being the lightest element, it has the highest place, followed by air, water, and finally the 'fourth,' crumbling earth, a multiple without any tendency toward unity.

Between these levels occurs a double movement: a movement of procession through which the power of the higher principle flows because of its overabundance toward the lower principle. A mirroring movement of conversion turns towards the higher principle, whose love of beauty is the real driving force. In his Enneads, Plotinus describes the descent or emanation of the universe from out of the One towards the creation of the sensory world as the procession which creates souls, the world, matter, and the multiple, while remaining One. The goal of spiritual life is the conversion or return and transformation towards producing ever-better organized forms, purer and more conscious, until they are reabsorbed back into the One.

In 'Rhythm of Wholeness,' Rudhyar brings the Midnight-Noon symbolic axis of the diurnal cycle (symbolically equivalent to the Capricorn-Cancer axis of annual cycle) closer to the Unity-Multiplicity polarity principles. (Fig. 32) In the arc of descent, between Midnight and Sunrise, when the principle of Unity is stronger than the principle of Multiplicity, emanation, or creative Word (Logos) becomes a *subjective* activity, where predominantly creative hierarchies develop archetypal foundations of the material universe.[151]

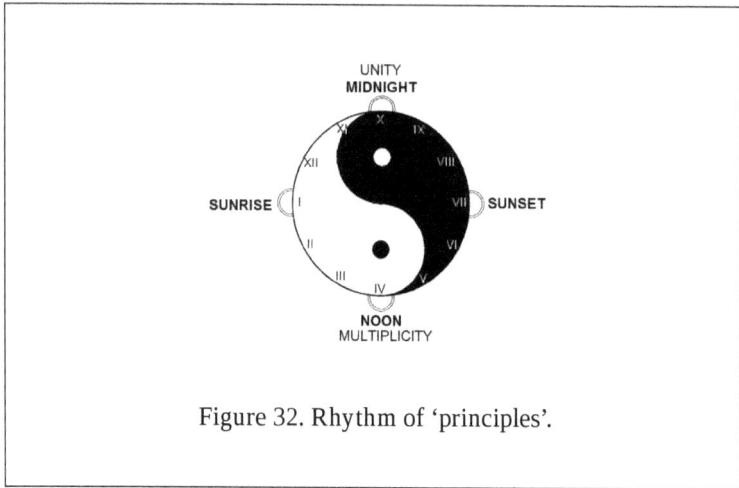

Figure 32. Rhythm of 'principles'.

From Sunrise to Noon, when the strength of the principle of Multiplicity is greater than that of the principle of Unity, the creative Word cyclically 'descends' and differentiates (involution) into specific 'Letters' which acquire an increasingly limited focus as *archetypes*. These principles, forms, and formulas of organization, progressively relate to and structure synchronously evolving (responding, also differentiating) material substances and systems.[152]

And, then, the arc of ascent, from the many towards the One etc.

Thus, the remarkable zodiacal language, based on the harmonious arrangement of the ternary and quaternary, is a model of formalization that can be used as a frame of reference in the contemporary scientific description of the universe. The parallel with the Yin-Yang symbol highlights the subtle asymmetry that enhances its power of integration both within inanimate matter and life.

ENDNOTES

139   Christopher McIntosh explains that the alchemical processes carried out by the Gold "[...] in which the material realm, although separated from the divine, was permeated by a divine element which could be refined out. This divine element was often referred to as the 'quintessence' to distinguish it from the four elements of air, fire, earth and water. It was the universal vital fluid, the breath that animated everything and was central to all alchemical operations, for this substance was a *sine qua non* for the making of alchemical medicines and for the preparation of the Philosophers' Stone used in the transmutation of metals." In Christopher McIntosh, *The Rose Cross and the Age of Reason* (Leiden & New York: Brill, 1992), 85.

140   Jean-Baptiste Morin, *Astrosynthesis: The Rational System of Horoscope Interpretation According to Morin de Villefranche*, trans. Lucy Little (New York, NY: Zoltan Mason Emerald Books, 1974).

141   Jung, *Synchronicity: An Acausal Connecting Principle*, § 987.

142   Jung to Hans Benders, April 10,1958, in *Letters 1951-1961, Vol.2*, 428.

143   Halberg et al., "Chronobiology's progress I," *Journal of Applied Biomedicine* 4 (2006): 1–38 http://jab.zsf.jcu.cz//4_1/halberg.pdf

144   Margaret Ostrowski-Sachs & C.G. Jung, *From conversations with C. G. Jung* (Küsnacht, Hornweg 28, ZH: C.-G.-Jung-Institut Zürich, 1977), 68.

145   Jung to Father Victor White, December 19, 1947 in *Letters 1906-1950, Vol. 1*. ed. G. Adler with A. Jaffé (Princeton, NJ: Princeton University Press, 1975), 54.

146   Ibid.

147   Jung, *Aion*, § 410. Following the 'Aion' publication, Pauli was struck by Jung's formula which led him to suggest an 'improvement' for their commonly agreed psychophysical quaternio of fig. 13. He expanded each term of the polarities into a (3+1) structure, e.g., time being the 'fourth' of 3-dimensional space. (See his letter to Jung of February 27, 1952 in Meier, ed., *The Pauli/Jung Letters*)

148   Dane Rudhyar, *The Pulse of Life* (Boulder, CO: Shamballa, 1978). Available online at http://www.khaldea.com/rudhyar/pofl/pofl_p1p1.shtml/

149   G. W. F. Hegel & Arnold V. Miller, *Hegel's Philosophy of Nature: Being Part Two of the Encyclopaedia of the Philosophical Sciences (1830)* (Oxford/New York: Oxford University Press, 2004), 26.

150   See Abellio, Raymond in *Dictionary of Gnosis & Western Esotericism*, ed. Wouter J. Hanegraaff et al. (Leiden, Netherlands: Koninklijke Brill NV, 2006),1–2, 1060–1061. Available online at http://www.antropologias.org/files/downloads/2012/12/123.pdf

151   Dane Rudhyar, *Rhythm of Wholeness, A Total Affirmation of Being* (Wheaton, Il: Quest Books, 1983). Available online at http://www.khaldea.com/rudhyar/rw/rw_c4_p2.shtml

152   Leyla Raël, *The Essential Rudhyar: An Outline and an Evocation* (Palo Alto, CA: Rudhyar Institute for Transpersonal Activity, 1983). Available online at http://www.khaldea.com/rudhyar/ess/ess_13.shtml

10. The Rhythmic Reflections of the Psychophysical Energy through a Diachronic Reading of the Zodiac

The universe emerges with a violent explosion of energy (Aries), is condensed into matter within the primordial nucleosynthesis (Taurus). After having interacted with its 'luminous' environment (Gemini), the universe becomes centered among the 'homes' of future galaxies (Cancer), shines and creates within the stars (Leo), transforms, improves, purifies its various forms within the biosphere (Virgo) and becomes conscious of itself within the Homo sapiens *(Libra)...And so on.*

The symbolism of the zodiac based on number-archetypes thus appears as a possible 'neutral' language sought by Pauli, allowing him to cut through psychophysical dualism. Is it possible that these traditional images, structured by numbers, could allow us to see deeper into the universal harmonies of the universe? One must indeed avoid any closed concordism that would add yet another analogy between a timeless symbolic system and the interpretation derived from a scientific model that could be refuted at any time.

This type of analogy would be very fragile. The challenge here is rather to support scientific effort in its eternal process of going to the very limits of reason while impelling these disciplines to re-consider ancient concepts that are more appropriate to the objects and phenomena considered and that could bring forth new meanings. Beyond this point, a metaphysics could take the relay as an imaginative source for new concepts. The zodiac may play the role of a symbolic 'interweaver' of events for the improvisational universe. As a vast web of potential interconnectedness, the zodiac offers the sciences a new light on old scientific problems. Its symbolic power can push them beyond phenomena, initiate new possibilities to grasp the very limits of mathematical formalization, and regain what had been eliminated in the classical conception of intelligibility: the return of the subject, the observer's consciousness, and the subjective and spiritual world of meaning.

Wolfgang Pauli has been one of the rare physicists to address the issue of the restoration of a unity between the researcher and the world around him. At the European Heritage Congress of Mainz in 1955, he delivered a lecture entitled "Science and Western Thought" in which he tackled the problem of the relationship between science and mysticism in a historical framework. He concluded:

> Taking a warning from the failure throughout the history of thought of all premature endeavours to achieve a unity, I shall not venture to make predictions about the future. As against the strict division of the activities of the human spirit into separate departments since the seventeenth century, I still regard the conceptual aim of overcoming the contrasts, an aim which includes a synthesis embracing the rational understanding as well as the mystic experience of one-ness, as the expressed or unspoken mythos of our own present age.[153]

Before returning to the 4 triads already glimpsed through the narrative of the universe's history, and broadening their meanings with the 12 zodiacal symbols, let us try to figure out how the alchemical axiom of Maria is expressed in the zodiac. How can the 'One is the same as the Four' be embedded in number $12 = 4 \times 3$? As far as the zodiac is an archetypal image of the Self, its main number is Four. The 12 number-archetypes, clad with the traditional images of Aries to Pisces should be understood with a primacy given to cardinal signs (Aries, Cancer, Libra, Capricorn) in a way similar to that given to the astrological angular houses of a chart. Obviously, such a primacy stems from the prevalence of the four directions of space and the four seasons in temperate countries. But ternary division of a quadrant is not the rule, various other ways of dividing up four-fold structures being observed in myths or dreams like Pauli's great vision of the world clock with its 4×8 divisions.

According to Jung's representations of the Self (Fig. 28), each quadrant whatever its partitions, is to relate to the whole pattern through self-similarity, i.e., perform the axiom of Maria on its own, towards oneness. In figure 28, the transformation process, represented by each grouping or tetrameria labeled with lowercase letters, reproduce the overall pattern of the dynamic process labeled with capital letters. This is a self-recursion pattern that Pauli (see note n°147) links to mathematical 'automorphism' and that Jung reflects in one sentence of his already quoted description, repeated here for convenience: "The 'sublimation', progress, or qualitative change consists in an unfolding of totality into *four parts four times*, which means nothing less than its becoming conscious." [emphasis added]

Coming back to the 4 triads detected in the history of the universe, their self-similarity to the whole pattern is equivalent to conceiving of them as incomplete (or as tetrads with missing fourths). The Three of the incomplete triad is taken as a unity and, relating back to the primal one, becomes the Four. This 'hidden fourth' may be interpreted "not so much to have 'originated' progressively, but to have retrospectively existed from the very beginning,"[154] in the 'life not lived' of the universe, that virtual state 'from which' a new quadrant will start. This state proceeds from the breaks caused by the four emblematic events (I, IV, VII, X) of the universe in the course of its becoming. For example, the radiative recombination of electrons with hydrogen atoms at about 380,000 years after the Big Bang, halts the uniform evolution toward the equilibrium of the thermal death into which the entire universe would have sunk, *had matter-light decoupling not occurred.* This state of uniform lethargy is the hidden One, 'before the one' of the second quadrant. The 'fourth' that makes 'the One being the same as the Four' is equivalent to the 'from what' a new quadrant will start. Briefly, Maria Prophetissa would see the 'fourths' of the four quadrants of cosmological evolution as follows: 'Between' XII and I (Fig. 33), the undifferentiated unity of symmetric or 'false' vacuum 'precedes' inflation of phase I. 'Between III and IV, the near-infinite uniformity of the universe precedes the growth of inhomogeneities in phase IV. 'Between' VI and VII, the uniform noise of proto-consciousness precedes the emergence of consciousness in phase VII. Finally, between IX and X, the hypothetical near-infinite negentropy of an Omega point à la Tipler 'precedes' the emanation of a new universe.

Let us now turn to the rich symbolic images of the zodiac that will broaden the meanings of each phase of the universe, leaving aside for chapter 11 the possibility to play with its rich pattern of relation that astrology calls 'aspects,' the main ones being oppositions and quadratures. A more detailed presentation of this parallel was done in the third chapter of our book

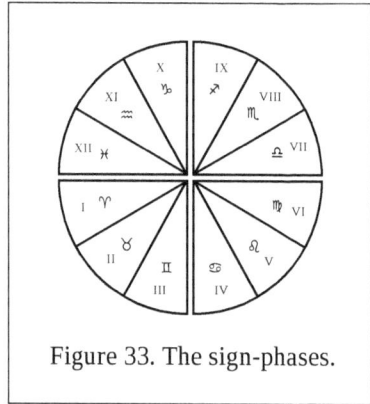

Figure 33. The sign-phases.

published in 1994.[155] Here, each sign-phase description will start with a summary of its symbolic significance which would be beneficial to complete with a parallel reading of the corresponding sign in, *Astrological Signs: The Pulse of Life.*[156] Then, each sign-phase will be followed by comments on possible archetypal reflections in the corresponding phase of the universe's history. (Fig. 33)

First Quadrant: Spring.

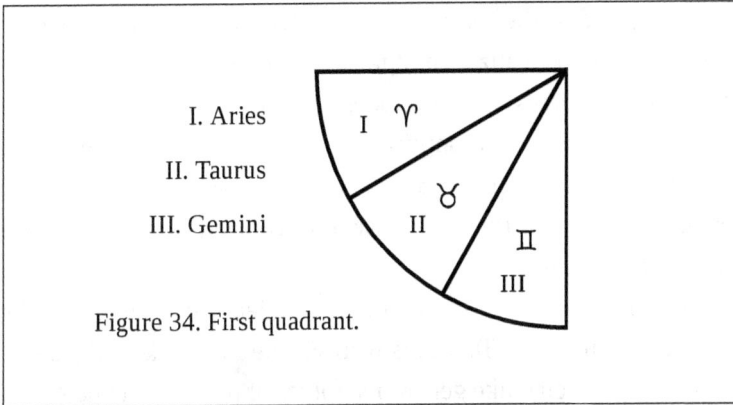

I. Aries

II. Taurus

III. Gemini

Figure 34. First quadrant.

I. Aries (♈), the fiery, male, and cardinal sign, is the starting point of the subject's identity. (Fig. 34) It corresponds to the burst of a primordial unity, the germ that pierces the seed. It represents the first stage of human personal development with an awareness of being a separate individual (in polarity with Libra which is the awareness of the non-ego).

This is the original impulse of the manifested being, the first breath of an individual life, the initial spark or the first emergence of an energy that has not yet taken shape. Out of all this, the need for freedom of action was born in order to distinguish itself from the environment and be aware of its sense of being as a separate existence.

From a contemporary cosmological point of view, the universe emerges from the quantum vacuum through a period of inflation when a phase transition occurs that sees the strong nuclear force detach from electromagnetic and weak nuclear forces. This phase of vertiginous expansion takes place in an extremely short time, which recalls the meaning of Aries in as much the universe differentiates from other potential universes (other vacuum fluctuations).

II. Taurus (♉), the earthy, feminine, and fixed sign, gives substance and depth to Aries' pulses. It corresponds to incarnation inside a form or a body and to the sense of permanence, stability, determination, perseverance and realization of talents. It marks the development of substance, resources and possessions encompassing innate abilities of body and psyche. The 'cosmic matter' is the medium through which generative forces coagulate, consolidate, and materialize. This energy of connection, attraction or repulsion, inertia and passive resistance sets the creative impulse.

From the perspective of modern scientific cosmology, the archetypal image of Taurus is well visible in inertia and passive resistance, which make generative forces of primordial nucleosynthesis coagulate. In this latter phase, the creative impulse of primordial inflation stabilizes during the cooling of the universe, which allows the creative surge to be clad with matter.

Visible matter forms through a slight predominance of matter over antimatter. Cohesion forces are involved in each point by confining quarks inside nucleons and anti-nucleons. Within minutes, the strong nuclear interaction bonds hydrogen and helium nuclei in a still undifferentiated universe. Only in the Leo phase will it resume its cohesion work in the hearts of stars, in a universe that has become 'individualized.'

III. Gemini (♊), the airy, male, and mutable sign, exteriorizes and interprets Aries' individual impulse once it has been channeled and materialized in Taurus. It corresponds to mobility development and to any type of exchange aiming to connect, communicate ideas, and build a dense web of close connections.

The twins symbolize the instinct of information research that is present within each child seeking to find how far he can go in all physical and psychological directions, before being stopped by someone or something. This contains a "fluid tie that circulates and twists, winds and creeps, hugs and tears before the next coming together. Gemini air seeps, imperceptible and intangible, through any fold, penetrating the smallest gap, so each interval becomes a place of contact."[157]

From the perspective of contemporary scientific cosmology, Gemini's archetypal image shows through the friction between matter and light, which sees free electrons of ionized matter constantly collide with photons. These particles of light analogically refer to the communication of information and to the need of integrating environmental impacts faced by a new-born when he reacts to impressions received in his daily contacts.[158] The friction of his skin with the environment develops an immediate capacity of awareness, which reminds us that "[...] the embryological origin for skin, brain and mind is the same. The ectoderm, our primary interface, is the outermost of the three primary germ layers of an embryo and the source of the epidermis, the nervous system, the eyes, and the ears, that is, interfaces."[159]

Second Quadrant: Summer.

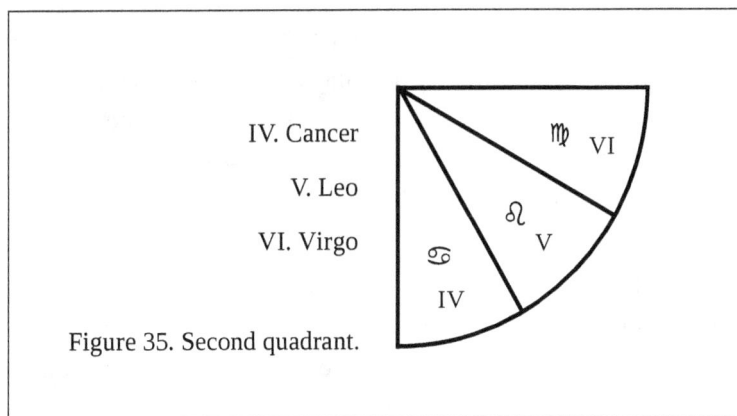

IV. Cancer

V. Leo

VI. Virgo

Figure 35. Second quadrant.

IV. The sign of Cancer (♋) beginning the summer solstice is a 'stopping point' of Solar progression traditionally designated as the 'door of men' which indicated the entrance through which souls descend. (Fig. 35) Here we are facing the entry into individual manifestation,

"[...] the embryonic setting in which are deposited the seeds corresponding, in the 'macrocosmic' order, to the Brahmānda or 'World Egg.' [...] The lunar sphere is properly the 'world of formation,' or the domain of the elaboration of forms in the subtle state, the starting-point of existence in individual mode."[160]

This watery, female, and cardinal sign, corresponds to anchoring firm roots, to construction of a concrete base of operation and emotional security. It signifies belonging to the founding matrix, home, family, and any self-protective fulcrum giving a sense of security.

From the perspective of contemporary scientific cosmology, the archetypal image of Cancer is reflected through the phase of hatching and growth of germs of stars and galaxies around gravitational nests. The appearance of specific forms corresponds to the expression of the base, the center of personality symbolized by Cancer. The motherly nature of molecular gas and dust clouds is often cited, such as the 'stellar nurseries' they keep deep within themselves.

V. Leo (♌), the fiery, male, and fixed sign, corresponds to the creative liberation of the inner vitality of being. Here is the personal expression as game, performance, or romance. It symbolizes the dramatic excitement and the development of a strong sense of individuality, courage, and self-confidence. With Leo, light is sparkling, day radiance is evident.

From the perspective of contemporary scientific cosmology, the archetypal image of Leo shows through the creation phase of heavy elements inside stars. Matter contraction causes a rise in temperature which, in the center of the protostar, triggers a thermonuclear fusion process that releases light and energy.

Here it is no longer the undifferentiated Aries fire, but that of individuation that burns and shines through individual forms. The antagonism between thermal pressure and gravity allows synthesized elements necessary for life such as carbon, nitrogen, oxygen, and phosphorus, which have been manufactured by successive generations of stars. While dying, the latter have, in fact, dispersed these biological precursors through space.

VI. Virgo (♍), the earthy, feminine, and mutable sign, corresponds to the development, improvement, and purification of the ego's creative expression. This is the stage of "the reversal by which existence diverts the ego's attention from the unique material reality and discovers the awareness of existence, which begins the ascent towards spirit."[161]

This is the sign of humility, discernment, self-criticism, and awareness of a lack, associated with growth crisis and inner doubts. The spirit that characterizes it, is separator, pragmatic, and meticulous while laboriously waiting for consciousness development.

From the perspective of modern scientific cosmology, Virgo's symbol reveals in the biosphere growth phase and selection of forms better and better adapted to the environment. Four billion years ago, the living organization emerges from a macromolecular whirlwind in storms and telluric convulsions. And, yet unable to replicate matter, a molecule that can make a copy of itself emerges, which characterizes life.

Mutations randomly occur and are selected for their survival value at a particular time. The 'natural selection' pressure seems to impose a purpose that drives an endless search for balance in a hostile world, generating increasingly complex beings. From first amino acids to *Homo sapiens*, it continuously improves forms and subtly refines behavior. Some 'preconscious' local life forms emerge within the cosmic 'soup.'

Third Quadrant: Fall

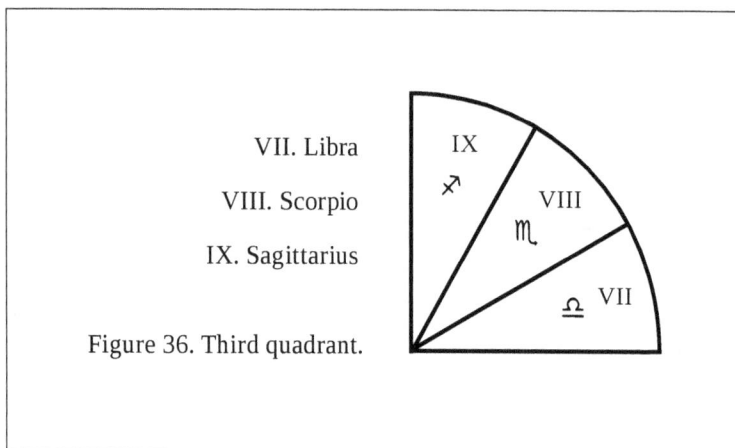

VII. Libra

VIII. Scorpio

IX. Sagittarius

Figure 36. Third quadrant.

VII. Libra (♎), the airy, male, and cardinal sign, corresponds to the emergence of reflective consciousness, a consciousness feeding back on itself that implies meeting others and being aware of non-ego; objectification. (Fig. 36) It is polar to the sign of Aries, which symbolizes how everyone becomes aware of one's uniqueness. Libra is a sign of balance, harmony, and equitable development.

From the perspective of contemporary scientific cosmology, the archetypal image of Libra appears in the emergence of reflective consciousness. Consciousness is like psychic substance associated with 'centers of consciousness' that have grown from phase IV seeds. The atoms of the eyes of the first *Homo sapiens*, shaped by generations of long gone stars, were struck by photons coming from other stars' atoms. One part of the universe 'spoke' to another part and thus, the universe became conscious of itself. Enormous accelerations experienced by a biological, cultural, and technological evolutionary process recall the opposite Phase I, the cosmic inflation that saw a dizzying expansion while creating space in an extremely short amount of time.

VIII. Scorpio (♏), watery, feminine, and fixed sign, corresponds to the ego surpassing in sexuality, transformation, and rebirth. This is a phase of intense renewal, regeneration, and personality stripping. There is confrontation between good and evil, light and darkness, because energy must transmute and sublimate in order to spiritualize itself. Here is the domain of Pluto who harasses each of the fallen beings to draw out the imperishable spark. Under its destructive fire, the last attachments to sensitive forms are annihilated.

From the perspective of modern scientific cosmology, Scorpio is reflected in the self-re-assessment of sentient beings in the evolutive universe. Planets detach from stars, and stars escape galaxies while losing any references to centers. Similarly, the 'egos' built in the Cancer phase and expressed in the Leo phase, see their limitations abolished in the Scorpio phase.

Since all matter is radioactive to various degrees, atomic nuclei eventually disintegrate to form other particles. This cosmological phase is symbolically close to Scorpio, where the need to unite with others manifests itself. In addition, conscious beings are challenged to transform and regenerate themselves, against stars that collapse into black holes and the disintegration of protons into lighter particles (electrons, neutrinos, photons). A denser energy form is to be renounced to release a subtler form.

IX. Sagittarius (♐), fiery, male, and mutable sign, corresponds to consciousness expansion along with its search for new fields of activity likely to involve the conquest of near and far environments. And, with its expansion in intellectual activities, thoughts or great ideas come together.

Sagittarius seeks to triumph over the continuing deterioration of natural energies and compression of the flow of energy so that it can release the light of thought and neutralize the tendency toward disintegration peculiar to all natural compounds. It represents the quest for eternal values and the quest for the absolute, or the search for distant connections that serve as the 'nervous system' for the social organism. More abstractly still, this sign corresponds to a "system of laws, ordinances, regulations which will enable the complex organism of society – the life of a city or nation – to operate satisfactorily."[162]

From the perspective of contemporary scientific cosmology, the archetypal image of Sagittarius shows through the distant future when black holes and normal matter have finally disappeared. From that moment on, conscious life will have to surpass and adapt its metabolism to post-material environment. It will have to extend throughout the universe and use the ultimate structures that remain, such as the electrons-positrons plasma.

As for Sagittarius, the universe of the distant future has a purpose that drives it to search for distant connections, a 'nervous system' for a social organism. Perhaps intelligent species of an open world will need to seek to survive despite the decays of positronium atoms supposed to occur through a final 'fireworks' of high-energy gamma photons.

These intelligent species will strive to eliminate cosmological horizons; namely, to make every region of the cosmos connect and perhaps – as Frank Tipler thinks – act in order to 'close' it. Indeed, this physicist assumes that, after the current acceleration phase, the universe will eventually shrink before life will have controlled half of the cosmos. Thus, the cosmological radiation that had fallen to very low temperatures will begin to warm up to reach the extreme temperatures of the early universe.

Intelligent species will continue to spread throughout the universe, adapting to an environment that has become very chaotic. For this purpose, they will use the gravitational field energy to process information which will be encoded in elementary particles appearing as the final singularity is approaching. This point, that Tipler calls the 'Omega point,' matches up with Sagittarius symbolism as the ultimate goal that drives him to approximate the realization of an intelligent, omnipresent Being.

Fourth Quadrant: Winter.

X. Capricorn (♑), sign starting the winter solstice, is a 'stopping point' of the solar progression once called 'door of the gods' implying that souls returned from earth towards heaven, where supra-individual states of being reside.[163] At Capricorn, the head and tail of the Ouroboros serpent, that unite past and future, re-connect. Capricorn is associated with the 'hidden God,' the unmanifested ultimate that periodically manifests by emanation of pre-cosmic consciousness. But it is also the realm of Kronos (Saturn) where eternity is manifested as time (Chronos). (Fig. 37)

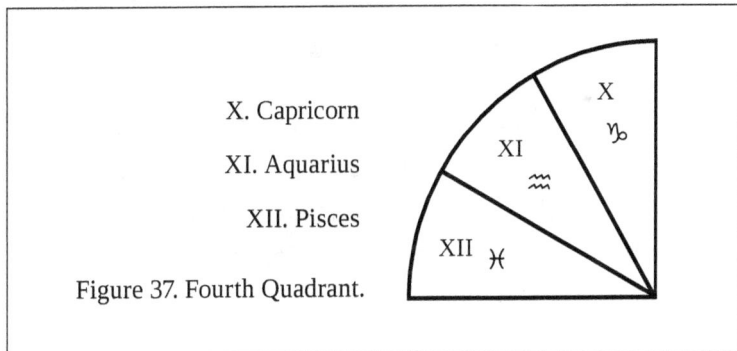

X. Capricorn

XI. Aquarius

XII. Pisces

Figure 37. Fourth Quadrant.

Capricorn, the earthy, cardinal, and female sign, represents the power of the creation of a permanent and stable basis for society. It symbolizes achievement, literally 'to come to a head,' – from Latin stem *caput* 'head' – or to the full realization as participation in the social or universal whole. This idea is translated in the institution, the State, the public authority firmly established by consensus, and any large organization oriented towards the fulfillment of the individual.

From the perspective of contemporary scientific cosmology, the archetypal image of Capricorn appears difficult to detect in the uncertainties of the universe's fate. Current measurements of the cosmic expansion rate imply its rapidly growing dilution, tending towards the quantum vacuum. This increasingly ethereal state corresponds to the Capricorn figure, emaciated by losses and privations which force him to always excel in accessing supreme vacuity.

According to Frank Tipler's hypothesis, once it appeared, intelligent life had to last forever, although this is incompatible with an accelerated expanding universe. He assumes that intelligent beings will gather it and concentrate as an 'Omega point' recalling Capricornian symbolism. In fact, at the final state – when the Omega point is reached – everything is known. The intelligence is total, it controls all existing forces and knows all that can be logically known. Tipler endows the 'Omega point' with omnipotence, omniscience, and omnipresent qualities that evoke divinity.

XI. Aquarius (♒), airy, male, and fixed sign, is where pre-cosmic consciousness emanating in Capricorn creates organization 'models' that will become the structural foundations of the universe. It is the breath of the collective soul, friends, groups, and clubs made up of complex associations of individuals, usually with idealistic views. Creation does not come from the individual as in Leo, but *through* the individual who participates in the community by injecting his progressive ideas and social visions.

From the perspective of modern scientific cosmology, in this state, 'prior' to the appearance of space, time, and matter, the Aquarian symbol emerges in the questioning of the origin of physical laws that are supposed to be identical to themselves for all eternity by the majority of physicists. It has been proposed that physical constants and large dimensionless numbers may change during the bounce of a cyclic universe. From this perspective, physical laws born from quantum information evoke quite well these new impulses specific to the Aquarian spirit.

The renewal and transforming ideal of this sign also appear in multiverse hypotheses. Aquarius is a power that acts as the creative vision of an infinity of virtual possibilities that would allow us to justify – through probabilities sets – the existence of such a peculiar universe as ours. In this phase, a vital impetus, an 'élan vital' would constantly generate universes in a powerful and unpredictable outpouring, reminiscent of inventions that lead to social progress related to the wintery sign.

XII. Pisces (♓), watery, feminine, and mutable sign, symbolizes the oceanic dimension of life; that which is conducive to imagination, to selfless service, to compassion, and empathy. Within this dimension it is easy to feel the energy of any environment. This stage of synthesis and uncertainty, of surpassing guilt and sidelining ghosts of the past, is a moment of achievement in understanding the end of cyclical process and preparation of the next cycle. Pisces is agitated by subtle emotional currents that prolong the spectrum of unresolved memories and anticipate the visions of a new world.

From the perspective of contemporary scientific cosmology, the archetypal image of Pisces represents the state of quantum vacuum, pre-existing and underlying the universe. A vacuum is an environment filled with virtual particles and antiparticles, constantly creating and destroying themselves, which is what bestows a vacuum with huge potential energy. "These virtual particles, ruled by limiting laws, like flying fish plunging in the liquid element, very quickly rediscover their natural environment, the sea of virtuality in this case, still called vacuum."[164]

As a perpetual agitated sea covered with foam, the quantum vacuum evokes Pisceans characterized by complete openness to flow from the unconscious. It is a sea of quantum superpositions not yet collapsed that may represent pre-consciousness 'fluctuations.' Vacuum fluctuations, which tend to make the universe uniform by deleting the initial irregularities, evoke the essential task of Pisces. It is to overcome negative attachment to memories of suffering and frustrations accumulated in the unconscious so as to let the spiritual seeds of the coming cycle germinate.

ENDNOTES

153   Pauli, "Science and Western Thought," in *Writings*, 147.

154   von Franz, *Number and Time*, 65.

155   Alain Negre, *Entre science et astrologie [Between Science and Astrology]* (Paris: L'Harmattan, 1994).

156   Rudhyar, *The Pulse of Life*, ibid.The 12 signs are symbolically associated with the 12 houses that depend on the diurnal cycle. The sign of Aries corresponds to the first house, the sign of Taurus corresponds to the second house, etc. See also Dane Rudhyar, *The Astrological Houses, The Spectrum of Individual Experience* (Garden City, NY: Doubleday, 1972). Available online at http://khaldea.com/rudhyar/astroarticles/problemsweallface_1.php

157   Véronique Lamboy, *L'art du Zodiaque* [The Art of Zodiac] (Chambéry, France: CREER éditions, 1982), 44. [translated by the author]

158   See House III, related to Gemini in Dane Rudhyar, *The Astrological Houses*, ibid.

159   Franco F. Orsucci & Nicoletta Sala, *Reflexing Interfaces: The Complex Coevolution of Information Technology* (Hershey, PA: Idea Group, 2008), xv.

160   René Guénon, *Symbols of Sacred Science*, trans. Henry D. Fohr (Hillsdale, NY: Sophia Perennis et Universalis, 2004).

161   Lamboy, ibid., 44.

162   Rudhyar, *The Pulse of Life*, ibid.

163   Guénon, ibid.

164   Michel Cassé, *Du vide et de la création [Primordial Void and Creation]* (Paris: Odile Jacob, 1993). [translated by the author]

11. Squaring the Circle and the Union of Opposites

'Aspects' and self-referential cosmology. Interchange of times and relational character of the zodiac. Heuristic richnesses of '4x3' matrix and enantiodromic cosmological 'formulas'.

11.1. Strange Loops and 'Aspects'

The evolution of the universe, based on the cosmological predominant model, reflects the numerical 'four×three' matrix that may be illustrated and enriched with the symbolic substance of the isomorphic structure of the zodiac. Astrology had been closely associated with the "as above, so below" of the Hermetic tradition where the microcosm of humanity mirrors the macrocosm of the universe. With the advent of modern science (that astrology had itself contributed to inspire at its onset), it has been contaminated in return by the mechanistic paradigm. Correlative links became causal links. But the power of its symbolic nature was not lost. The imaginative logic of its diagrammatic representation reappears through a new approach to the cosmos. The alleged causal influence exerted upon us has been changed into a genealogy link: we are the "children of the stars." And this genealogy link is reflected through the mandala of the zodiac.

The zodiac circle naturally highlights 'aspects' that are the angular – multiples of 30° – relations between signs. In the delineation of a particular chart, oppositions and squares are the two main 'aspects' that highlight challenges to the individuation process: ego fulfills and gives way to the Self, symbolized by the circle-squaring that unites opposites in cross linkages of shadow and light, in a synthesis of personality.

Transposed in terms of the cosmological time of the universe, these very same 'aspects' connect as strange loops some of its past events to its present and future. This entanglement evokes John Wheeler's self-referential cosmology, for which the past only exists from the moment it is recorded in the present. (Fig. 38)

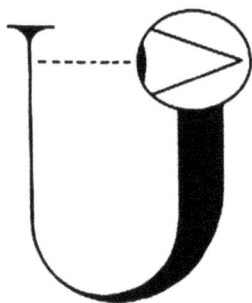

Figure 38. Wheeler's
self-referential cosmology.
*(Image reprinted with permission
from John A. Wheeler.)*

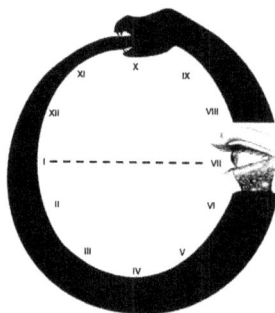

Figure 39. The alchemical
symbol of Ouroboros
reflecting the transformation
of consciousness.

In quantum mechanics, based on the necessity for the observer to reduce the wave function of a particle, Wheeler has developed the idea of a participatory universe where observers are necessary to bring the universe into existence. Feedback loops can thus represent the participants' contributions to the permanent creation of not only the present and the future, but also the past. Building on the zodiac reading grid (Fig. 39), we will consider what is implied by this interactivity of certain events in the universe at the most fundamental level.

11.2. The Interchange of Time

The narrative of the evolving universe has thus appeared to resonate with a diagrammatic, punctuated by the quality of first integers and symbolically enriched by zodiac images. It could thus reflect the manifestation of a movement of circumambulation of a unitary reality, a psychophysical reality that is potential, relative to our empirical world.

The zodiac is just one example among many other traditional numeric structures based on the first four integers, such as the Pythagorean tetractys, which Edward Edinger used to show correspondence with the Sefirotic Tree.[165] (Fig. 40) In both symbols, the initial descent of the sequence 1,2,3,4 represents the involution process from a state of original wholeness to the fourth stage of the 'individual fully living in the world.' Then, following the inadequacy of that state of being, there is the task of individuation that mirrors in the ascending sequence 4,3,2,1 with the three steps of transition corresponding to the 'three stages of the coniunctio.'

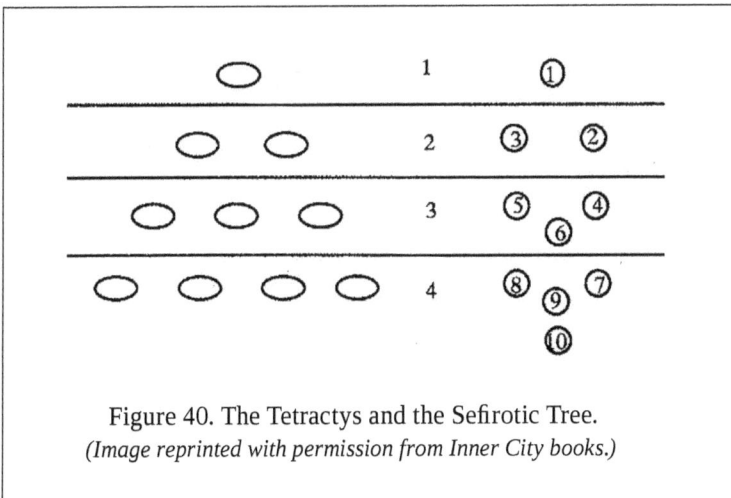

Figure 40. The Tetractys and the Sefirotic Tree.
(Image reprinted with permission from Inner City books.)

However, the zodiac wheel is the one which accumulates the most symbolo-geometric characteristics. The principal one is its circular nature, which is traditionally the figure of cyclical time. While harmoniously interlacing Three and Four, the zodiacal circle somehow solves the mystery of the transition between these two numbers, so often highlighted since the time of Mary the prophetess.

In addition, the circle may be considered as a reduction of the sphere in an equatorial plane. The sphere is the representation comprising most symmetries. As already mentioned, the sphere 'thickens' the zodiac with involutive, devolutive, and evolutive processes. In terms of various conceptions of time, it is an 'interchange' where three major temporal modalities engage. (Fig. 41)

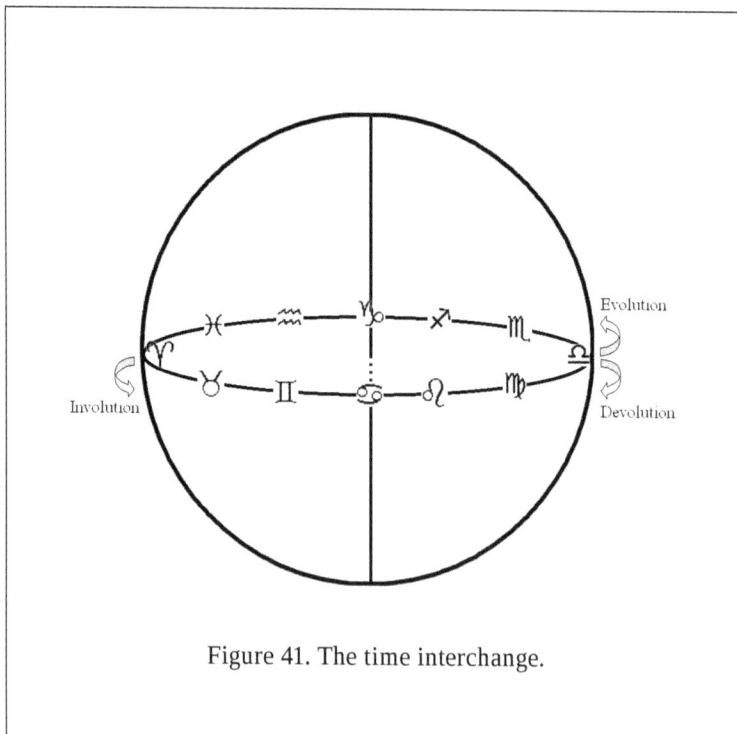

Figure 41. The time interchange.

Symbolically, the center of the sphere refers to an instance that reflects the extratemporal. It is a potential time at the deepest level of reality, where lies the *eternal substance* of absolute or transcendental 'no-time.' We are facing the acausal and timeless time in the sense of Kairos, that of synchronicities studied by Jung and Pauli: that which reveals meaning. It may be related to the reality behind the emergence of experimentally observed quantum entanglement phenomena, or to the effect of global on local that seems to be implied by the sensitivity of the Foucault pendulum oscillation plane to the most distant galactic objects.[166]

These two non-separability phenomena, whose nature remains unclear, refer to the conception assigned by Jung to the unconscious: "It seems to me like an omnipresent continuum, an unextended Everywhere. In other words, when something happens here at point A which touches upon or affects the collective unconscious, it has happened everywhere."[167] In deepest agreement with the physical structures, this non-time would then be a form-giving nothingness; a condition of possibility for other times, traditionally associated with figures such as Chronos (linear time) and Aion (cyclic time or eternity). These two time 'personifications' seem to appear in the interpretation of the contemporary cosmological predominant model.

Through reduction of the spherical form, these two temporalities appear as geometrical projections. The linear aspect of the flow of becoming is projected onto the vertical polar axis, and the cyclical aspect onto the equatorial plane. The 6 polarities, associated with the 12 signs-phases of the story, thus work together in time consciousness through a double-torsion movement around the vertical polar axis. On this axis, two spirals are entangled: the upward spiral engages in odd or masculine polarities phases (I-VII, III-IX, XI-V) corresponding to involutionary and evolutionary growth of information or negentropy. Conversely, the downward spiral shows the trajectory of pairs or feminine polarities phases (II-VIII, X-IV, VI-XII) which correspond to an increase in 'devolutive' entropy.

11.3. The Three Crosses of the Relational Physics of the Universe

Leibniz was opposed to the conception of absolute space and time
à la Newton, and argued that these were, *like numbers*, only a
general ordering principle between things. In his *Monadology*, he
stresses on the relations that each being has with all other things:

> [...] This connection or adaptation of all created things
> to each and of each to all, means that each simple sub-
> stance has relations which express all the others, and,
> consequently, that it is a *perpetual living mirror of the
> universe.* [...] All is a plenum (and thus all matter is
> connected), and in the plenum every motion has an effect
> upon distant bodies in proportion to their distance, so
> that each body is not only affected by those which are
> in contact with it and in some way feels the effect of
> everything that happens to them, but is also mediately
> affected by bodies adjoining those with which it itself
> is in immediate contact. Wherefore it follows that this
> inter-communication of things extends to any distance,
> however great. And, consequently, everyone feels the
> effect of all that takes place in the universe, so that he
> who sees all might read in each what is happening
> everywhere, and even what has happened or shall happen,
> observing in the present that which is far off as well in
> time as in place [...].[168]

Today some physicists step anew towards this conception. They
attempt to *describe the evolution of the world without resorting to
time,* by simply eliminating it from their theories.

This relational character is potentially embedded inside the
zodiacal diagram and may bring to mind qualitative and/or quantitative relationships between events, possibly suggesting invariants
or new universal structural constants. While approaching the limits
of explanations, the zodiac could then be used to resonate and to

reason, a sort of 'relational-conceptual' laboratory for questioning possible links between various cosmological events.

Oppositions and squares are the two main aspects, drawing three crosses that connect signs of the same quality. The cardinal cross (Fig. 42) connects the first signs of each quadrant; the fixed cross (Fig. 43) connects the second signs; and the mutable cross (Fig. 44) connects the third ones. Analogically transposed from the zodiac to the universe's history, the three crosses invite a relational reading of the universe.

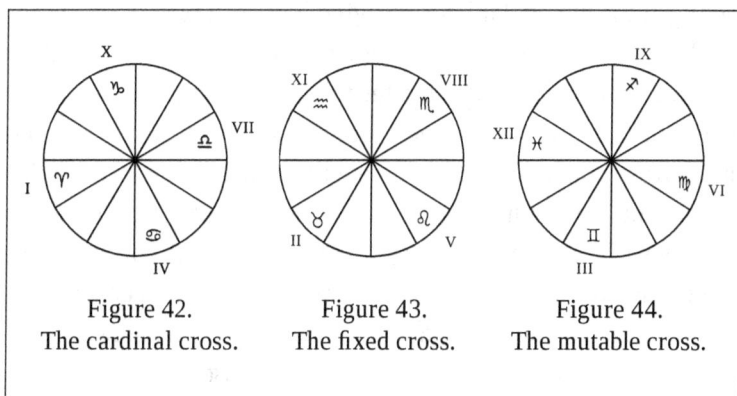

Figure 42.
The cardinal cross.

Figure 43.
The fixed cross.

Figure 44.
The mutable cross.

The Mutable Cross.

Phases of the evolving universe associated with Gemini, Virgo, Sagittarius, and Pisces bring electromagnetic forces into play. (Fig. 45) These forces and their conveyor particles (photons) are responsible for cohesion between nuclei and electrons. Mutable signs are related to cosmological transition and change phenomena, where information is transmitted through photons.

While moving between brain neurons, information appears related

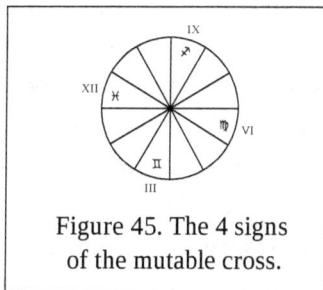

Figure 45. The 4 signs
of the mutable cross.

to the development and transmission of consciousness, hence the link between the electromagnetic field and the brain. We shall return to Roger Penrose's theory of the origin of consciousness, which gives a fundamental role to the gravitational field. However, as far as the transmission of nervous impulses are concerned, they are complex electrochemical interactions occurring within the human brain and clearly involve the electromagnetic field.

In Phase VI (Virgo), electromagnetic interaction forms molecules that are the basis of all chemical and biological processes acting at the molecular level of living beings. It plays a role in the evolution of the biosphere in which 'proto-conscious' entities – ranging from particles of matter to molecules of life – play the game of 'selection' and where proto-consciousness fluctuations jump from one living species to another. Such or such adaptation tool to the environment, such or such means of struggle for life, such or such acquired characteristics disappear with species, while alternative improved and selected traits or behaviors emerge.

It has been known for a long time that electromagnetic information travels in a vacuum. The vacuum, associated with phase XII (Pisces), is full of quantum fluctuations that communicate information throughout the universe.

Amplified by cosmic inflation, seeds of matter from which galaxies germinate are present in the light-matter plasma of Phase III (Gemini). The electromagnetic force continuously causes matter and light to interact. This plasma tends to organize itself around matter overdensities, but photons – striving to link negative electron to positive nucleus in order to form hydrogen atoms – are still too energetic throughout this friction phase between light and matter. The overdensities will be able to grow only from the decoupling following friction.

In the deep post-material future of phase IX (Sagittarius), electromagnetic interaction with infinite range connects electrons to positrons into positronium atoms. If the universe expansion rate is not too large, intelligent species encoded on these atoms should strive to suppress cosmological horizons. To this end, they would build a unified being, before positron/electron fusion, similar to matter/antimatter annihilation supposed to have taken place in the first moments of the universe.

The Cardinal Cross.

Gravitational force causes a mutual attraction between uncharged material bodies and appears to be involved in the 4 phases associated with the cardinal signs: I (Aries), IV (Cancer), VII (Libra), and X (Capricorn). (Fig. 46) Attraction is obvious in Phase IV (Cancer) when, following light/matter decoupling, gravity makes density inhomogeneities grow. It is less obvious in phase X (Capricorn), except in the case of a closed universe where triumphant gravity makes the universe fall back into a *Big Crunch*. The conception of gravity as an entropic force may one day allow for multiple planned scenarios coming together into one single event.[169]

As mentioned above, Penrose's falsifiable theory has been associated with events from phases I and VII through a triggering phenomenon associated with gravity. It provides an objective physical threshold, yielding a plausible lifetime for quantum-superposed states.[170] Each

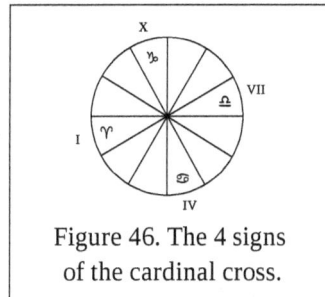

Figure 46. The 4 signs of the cardinal cross.

wave-function reduction event is a purely physical process, a primitive kind of 'observation,' a moment of 'proto-conscious experience.' Reaching this gravity-related objective physical threshold would cause moments of consciousness (VII) in the brain. In the same way, while dealing with the cosmological theory, Paola Zizzi has shown that reaching the same kind of threshold would cause the collapse, at the end of inflation (I), of a state of quantum superposition of multiple universes towards the classical world in which we live.[171]

These hypotheses, connected to measurement theories of quantum mechanics and to space-time geometry, suggest the existence of a link between biomolecular processes of the brain and the basic structure of the universe. It may be surprising to see such a high priority given to gravitational interactions in VII, as they are much weaker than electromagnetic or Van der Waal's forces, not to mention the damping of the quantum effects by thermal noise that exists in the warm environment of the brain. But in this extension of quantum physics proposed by Penrose, gravity has an effect only on the *uncertainty* of energy, not on the energy itself. This uncertainty of the energy is converted into 'self-energy' (E_G), which is the *difference* between mass distributions of quantum states, which is to say, the energy required to move a component of the superposition into the gravitational field of the other. "This energy uncertainty is the key ingredient of the computation of the reduction time τ, and it is appropriate that this energy uncertainty is indeed far smaller than the energies that are normally under consideration with regard to chemical energy balances, for example. If it were not so, then there would be a danger of *conflict* with normal considerations of energy balance. Nevertheless, the extreme weakness of gravity tells us there must be a considerable amount of material involved in the coherent mass displacement between superposed structures in order that τ can be small enough to be playing its necessary role in the relevant OR processes in the brain."[172]

The Fixed Cross.

Events associated with phases II (Taurus), V (Leo), VIII (Scorpio), and perhaps XI (Aquarius) involve strong and/or weak interactions that allow just as much particle attraction inside atomic nuclei as some nuclear decays such as β disintegration. Strong interaction allows for the cohesion of nucleons, and thus concentrates matter in phase II. In phase V, it locally creates heavy nuclei inside stars. (Fig. 47)

The weak nuclear force – neither attractive nor repulsive – is not responsible for any known structure in the universe. However, it is the only one of the four forces to discriminate between right and left, and could play a crucial role in phase II in matter's predominance over antimatter.

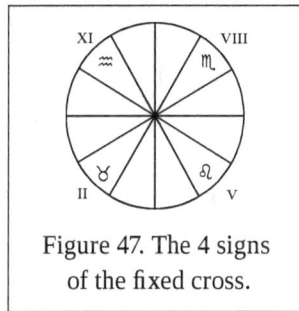

Figure 47. The 4 signs of the fixed cross.

In Phase V, the weak nuclear force helps in making stars shine by participating in thermonuclear reactions, while in Phase VIII it transmutes matter by disintegrating heavy particles into their 'ground state.' Thus, it does operate in the proton decay process. Its role in phase XI becomes indistinguishable from the other three interactions.

11.4. *Coniunctio Oppositorum:* The Six Oppositions

Within the holistic relations field represented by the zodiacal symbolism, we will now turn our attention to the six 'opposition aspects' that illustrate the *'conjunction of opposites.'* This is a widespread pattern observed in the mythical world as well as in the Beatles' lyrics:

> *"The deeper you go the higher you fly, The higher you fly the deeper you go, So come on! [...] Your inside is out and your outside is in, Your outside is in and your inside is out."*[173]

Opposites that emerge from each other and exist with respect to the other, are never definitively eliminated. They consist, interpenetrate, and mutually engender one another.

Would there be, at the historical scale of the universe, the equivalent of moderation laws like chemical, biochemical, or electromagnetic equilibria? Delays, counted in billion of years would occur between actions and reactions of these cosmological enantiodromic processes. Out of the 6 oppositions, the two most important ones belong to the cardinal cross: I-VII and IV-X. (Fig. 48) The four additional oppositions, fixed and mutable crosses, merely deploy antagonisms that are born from equinoxial broken symmetries (I-VII) or from solstitial attractions, (IV-X) by first substantiating them (fixed quality), then synthesizing them (mutable quality).

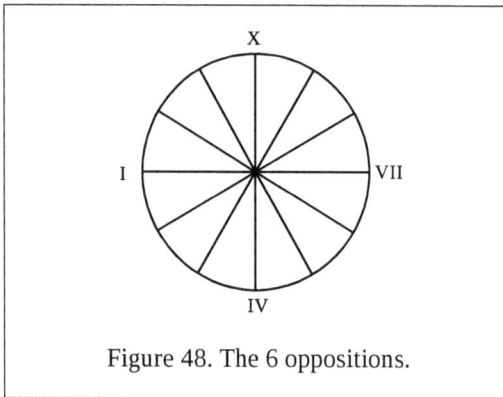

Figure 48. The 6 oppositions.

I-VII Polarity or Aries-Libra

Phase VII is the reflective consciousness occurring in an exponential trend, in time as well as in space. At its opposite, the cosmic inflation of Phase I is a theory which tries to explain the creation of space out of a quantum vacuum, with an exponential rate as well. What a mysterious notion space is: physical space created by inflation in I appears in polarity with the creation of 'psychic space' in VII, such as space awareness that develops in the psychological interior of individuals.

As previously mentioned, a possible enantiodromic link between I and VII (Fig. 49) was pointed by Paola Zizzi. This astrophysicist considers the universe in terms of quantum information: at the beginning of inflation (phase I), the universe was in a superposition state of quantum registers.[174]

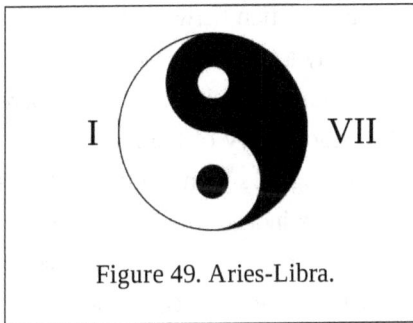

Figure 49. Aries-Libra.

Using the Penrose formula to calculate the critical threshold corresponding to quantum collapse, she found that the end of inflation ($\sim 10^{-33}$ second) coincides with the threshold causing the collapse of the quanto-relativistic wave function that linked multiple universes. This collapse thus marks the reduction of the virtual universes to our universe. This process is accompanied by a cosmic event – a *moment of conscious awareness* – in which individual consciousness would be microcosms. According to Zizzi's calculations, there would have been 10^9 quantum superpositions of multiple universes before 'objective reduction' triggering which marks the end of the inflation era.

This number, Paola Zizzi recalls, refers to the same number calculated for each moment of human consciousness. The human brain is indeed characterized by a state where 10^9 tubulins are in quantum superposition states before being subject to quantum-gravity collapse when a 'moment of consciousness' occurs.

The work of Penrose, Hameroff, and Zizzi thus merge with Arthur Eddington's intuition that "the substance of the world is mental."[175] The 'moment of awareness' or the *Big Wow* of the end of inflation recalls the Jungian cosmic unconscious generating time and space. It is an unconscious which is also a "[...] relation field where consciousness arises as a movement of negation *at the same time necessary and predetermined in its form*," that Michel Cazenave calls "a dialectical negation, which connects to itself."[176]

The enantiodromic relation between I and VII, rediscovered by Paola Zizzi, would imply a proto-consciousness in the form of information that exists from inflation. It would grow increasingly clearer throughout the rhythmic history of the universe, up until a kind of pre-consciousness in biological systems of Phase VI. In this latter phase, Stuart Hameroff suspects indeed the presence of the phenomena of 'objective reduction' behind the Cambrian explosion of small worms and sea urchins that suddenly appeared half a billion years ago. Building itself gradually through successive dialectical negation of the universe's evolution, the proto-consciousness of Phase I would thus rush towards its opposite, the reflexive consciousness of Phase VII.

Coming from an archetypal ground, reflective consciousness, at the same time connective and separative,

> [...] responds to its *idea* while, nevertheless, distinguishing itself, from this formative unconscious and, developing its principle which is that of separateness, that of clear and neat distinction, that of putting itself in conflict with its own matrix: "The hero's main feat is to overcome the monster of darkness. It is the long-hoped-for and expected triumph of consciousness over the unconscious. Day and light are synonyms for consciousness, night and dark for the unconscious [...]. 'And God said: Let there be light!' [These words] are the projection of that immemorial experience of the separation of consciousness from the unconscious."[177]

II-VIII Polarity or Taurus-Scorpio

With expansion, the Phase II universe cools, revealing visible matter that forms by the slight predominance of matter over anti-matter. This dominance is perhaps due to the electroweak phase transition that occurred at the temperature corresponding to an energy of 100 GeV (presently out of range of an experimental verification). During this phase, particles acquire their *mass* which is a property forecast by the recently tested Higgs mechanism (2012). When energy drops below 150 MeV, the strong interaction bonds quarks that become confined within nuclei.

Thus, both nuclear forces are involved in Phase II to coagulate energy coming from inflation. But only the weak nuclear interaction, responsible for beta decay, will continue its purifying role. In fact, the various constituents of matter are radioactive to varying degrees and – whether at the particle level or at the level of atoms, molecules, planets, stars, galaxies – the material substrate born in Phase II carries with it the seeds of its future total disappearance that will occur in phase VIII. (Fig. 50)

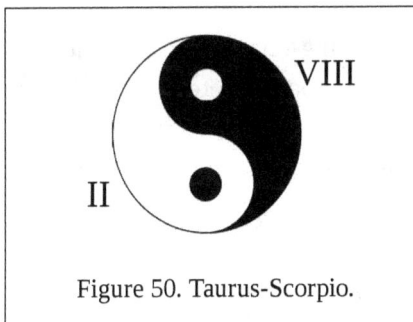

Figure 50. Taurus-Scorpio.

In phase VIII, the weak force transmutes heavy particles toward their 'ground state' through the very long decaying proton process. Various theories predict its disappearance, but present measurements only offer an experimental lower limit of 10^{33} years. Some quantum gravity theories predict that protons would end up disappearing before 10^{50} years. Photons, neutrinos, electrons and positrons would be the only remnants.

The II-VIII polarity suggests that these particles are the 'material' mirror image of an 'informational x', a cosmological 'entity' built from the reflexive consciousness that would face the 'unconscious' matter of Phase II. As far as an analog to Le Chatelier's moderation principle would hold, the chemical-like enantiodromic 'formula' would read:

Matter + Antimatter ↔ 'Superconscious' Positronium + Neutrinos + Light

III-IX Polarity or Gemini-Sagittarius

In this polarity two plasmas face one another:

- that of phase IX is a post-material plasma in which only remain photons and supposedly stable particles such as neutrinos, electrons, and positrons. Coming from an ultimate transformation of matter, this plasma is subjected to an accelerated expansion, succeeding to proton decay and dissolution of galactic black holes. The electromagnetic interaction retains negative electrons and positive positrons in the structures of positronium atoms. Speculative theories (Tipler) see these particles and antiparticles as constitutive of the Being-universe. Furthermore, the theories give them the ability to 'shape' the fate of the universe.

- that of phase III is an opaque plasma where photons – intimately mixed with elementary particles of matter – are constantly absorbed and re-emitted by electrons. Accordingly, light cannot rouse itself from matter. This matter/light friction only stops at the end of 380,000 years when the electromagnetic interaction, conveyed with cooled photons, manages to form hydrogen atoms. Only at that time can light and matter be separated, and light will be free to move in a universe that has become transparent. (Fig. 51)

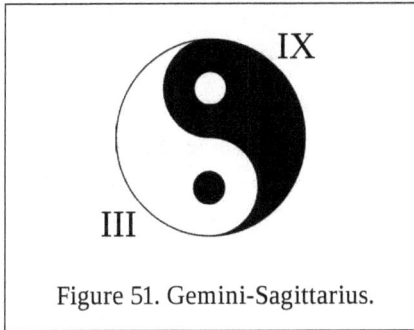

Figure 51. Gemini-Sagittarius.

An enantiodromic interpretation of III-IX polarity could be stated as follows: the density of matter inhomogeneities of Phase III, unable to individualize for being mired in light, locally 'struggle' in order that a light-matter communicative duality can take place in every point of the universe.

These germs or seed galaxies, potential centers of consciousness, will grow and backfire midway through the cycle in phase IX, into a united 'poem' of nearly pure light. The infinitely slowed union of extremely spatially diluted positrons and electrons will then constitute a positronium plasma. This distant relative of hydrogen will be very unstable as both of its components are annihilated into final fireworks.

The III-IX polarity is a 'mutable' polarity; that is, a transition into something else. On side IX, it 'prepares' the 'Archer Sagittarius' universe aiming at the Omega point target. On the other side, the Phase III computer universe 'calculates' the right initial amplitudes for galaxy seeds that will allow for the growth of complexity.

Missing the target in phase IX refers to the relative failure of phase III 'heat death' universes that would not have the 'right' constants of physics. What awaits these universes is symbolically linked to the *'end of things,'* which is one of the meanings associated with phase IV by astrological tradition.[178] On a purely cosmological level, galaxy germs cannot hatch and these expanding while cooling universes indefinitely deteriorate towards the absolute eternal cold of diffuse space background.

IV-X Polarity or Cancer-Capricorn.

Phase X events depend on the proportions of energies that fill the universe and on the proposed capacity bestowed on the conscious beings who are supposed to be able to force the ultimate fate of the universe. The white dot on the black background (or Yang germ over Yin), corresponds to the awakening of Day-Force that individualizes and differentiates. It symbolizes the lack of a fully achieved unity, a sort of imperfect Omega point à la Tipler: not completely omnipresent, omniscient, or omnipotent. (Fig. 52)

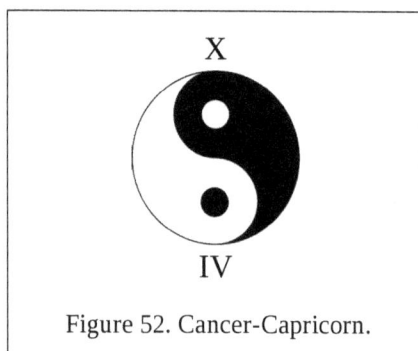

Figure 52. Cancer-Capricorn.

Would there be an enantiodromic process connecting this imperfect unity-seed of phase X to Phase IV inhomogeneities of matter, which are seeds of future galaxies and other cosmic structures? In other words, would imperfect unity in X transform, through a reversal mid-cycle into an imperfect multiplicity in IV? Or, starting from IV, would multiple material overdensities turn, mid-cycle in X as a 'light seed?'

The IV-X polarity is also that of the dialectic One (X) and Multiple (IV) which may be expressed using black hole physics. Still largely theoretical, quantum treatment of this concept would imply, according to Hawking, that black hole and white hole be interchangeable; one being the symmetrical model of the other by a simple reversal of the direction of time.[179] While crossing X, the black hole as a final synthesis of end-of-a-universe black holes would become the 'initial' white hole of a new universe. Through an enantiodromic reversal, it would re-appear in IV as the multiple germs of galaxies that are multiple *potential* black holes. Conversely, the future myriad creative potential stars in IV, bearers of life and consciousness, would end up as a unique black hole, evaporating and bouncing in X as a pre-temporal white hole from 'before' a new universe.

V-XI Polarity or Leo-Aquarius.

If strong nuclear force wakes up in phase V, locally creating heavy elements necessary for complexity growth, at opposite phase XI – heart of quantum-gravity field – many physicists suspect the existence of a process connected to the creation of laws and the constants of physics.

As a mirror image of the antagonistic process within phase V stars, the overall symmetry invites us to see in phase XI the vacuum energy (or part of it in the form of dark energy) with an anti-gravitational behavior, repulsive vis-à-vis the attractive gravity of phase X. The creation of information associated with the laws and the constants of physics could emerge in XI as virtual universes that build by antagonism with the germinal Phase X entropic[180] universe. (Fig 53)

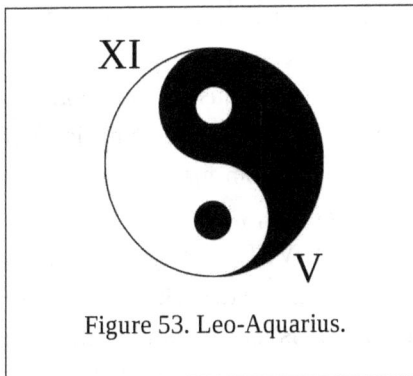

Figure 53. Leo-Aquarius.

Phase XI could be related to infinite quantum superpositions that linger in unobserved states. Through a mid-cycle swing in phase V, these virtual mathematical universes – to which may be added all other possible types of universes that may arise on the way like in eternal inflation – would gradually *withdraw*, making room for the enantiodromic process of a full actualization of creation within a *single* material universe, namely the basic elements necessary for increasing complexity. This enantiodromic creative process would unfold a unique sensory universe, unveiling the intelligible multiverse while hiding it behind the veil of matter.

The fixed cross seems to be involved in different aspects of energy and (visible and black) matter. Samy Maroun and Carlo Rovelli[181] have hypothesized that dark energy is linked to quantum properties of space-time and can be seen as a manifestation of quantum fluctuations of the vacuum that take place as long as the quantum state of space-time has not decohered, which would correspond here to phase XI. As for dark matter, they suspect the influence of *decohered* particles[182] in space-time close to galaxies.

In other words, could dark matter, whose presence around galaxies is suspected (V), be an 'intermediate state' between dark energy (XI) and ordinary matter (II)? In line with the work of Penrose-Hameroff on the quantum origin of consciousness, Samy Maroun and Carlo Rovelli suggest that remote correlations (EPR) of entangled particles "could be linked to the existence of coherent states and thus, ultimately, to the existence of a psyche intrinsic to Dark Energy."[183]

As they themselves note to justify their bold assumptions, they go as far as possible in their research in physics without exceeding the limits that separate it from metaphysics. On the basis of the numeric symbolic structure, we suggest that the 'informational x' added to balance the II-VIII polarity (discussed above) may be connected to a psyche related to dark energy. In VIII, this psyche would be 'superconscious' since coming after the emergence of consciousness.

VI-XII Polarity or Virgo-Pisces

Phase XII could be the synthesis between phase X attractive gravitation and antagonistic phase XI, where vacuum pressure suspends virtual universes in quantum superposition states, bound by the quanto-relativist wave function. In phase XII, vacuum fluctuations, in resonance with curvature fluctuations of space-time, could correspond to an effacement process whose duration would be of the order of Planck time. (Fig. 54)

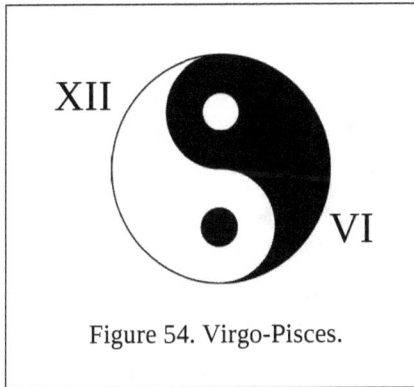

Figure 54. Virgo-Pisces.

For the universe as for the individual, forgetting is a necessary function that clears memory by removing superfluous information. In physics, it is now well established that this irreversible elimination process leads to an increase in entropy.[184] The phase XII universe is then a symmetrical empty space, a 'false vacuum' endowed with huge potential energy. Uncollapsed quantum superpositions are like the foam of the waves of an ocean of energy formed by the irreversible erasure of potential universes.

This erasure in XII could be the enantiodromic reversal of an irreversible process of 'selection' that, in phase VI, leads to the emergence of life. Some research in this area involves the same processes used to explain the emergence of consciousness in tubulin and microtubules of the brain. Quantum coherence has been observed in photosynthetic systems of the biosphere. Objective reduction (OR) events would have taken place in both plant photosynthesis and eukaryotic cells, which appeared 1.3 billion years ago.

First, these events would have occurred without orchestration, then they would have been more orchestrated throughout evolution. In this respect, Hameroff suggests that primitive consciousness caused by Orch OR would have existed in animal species, and that this phenomenon would have precipitated the accelerated evolution of the Cambrian explosion 540 million years ago.[185]

ENDNOTES

165   Edward F. Edinger, *The Mysterium Lectures: A Journey through C.G. Jung's Mysterium Coniunctionis*. (Toronto, Canada: Inner City Books, 1995), 282.

166   Michel Cazenave, "Synchronicité, Physique et Biologie," ibid., 52. Austrian physicist Ernst Mach (1838-1916) who, besides, happened to be Pauli's godfather, assumed that a body's inertial mass is determined solely by other matter or energy in the universe – a kind of ubiquity of the universe as a whole.

167   Jung to Dr Albert Oeri, January 4, 1929, in *Letters 1906-1950, Vol.1*.

168   G.W. Leibniz, *Discourse on Metaphysics and Other Writings*, ed. Peter Loptson (Peterborough, Ontario: Broadview Press, 2012), § 56 and §61.

169   See note n°131.

170   This lifetime (or decoherence time) τ is related to the gravitational 'self-energy' E_G (explained in the following paragraph) by the formula: $\tau = \hbar / E_G$. The reduced Planck constant $\hbar$ (or angular momentum unit - amu) was introduced by Dirac to describe the spin of the electron: $\hbar = h / 2 \pi \approx 1{,}054\,571\,800 \times 10^{-34}$ J·s

171   Paola Zizzi, *Emergent Consciousness: From the Early Universe to our Mind*. Available at: arXiv:gr-qc/0007006

172   Hameroff & Penrose, ibid., 559.

173   J. Lennon & P. McCartney, "Everybody's Got Something to Hide Except Me and My Monkey" [Recorded by The Beatles]. On *"The Beatles" (White Album)*. London: EMI, 1968.

174   Paola Zizzi, ibid.

175   A. S. Eddington, *The Nature of the Physical World* (Cambridge, UK: Cambridge University Press, 2012).

176   Cazenave, *La Science et l'âme du monde*, 51.

177   Ibid., 57, who quotes Jung's *The Archetypes and the Collective Unconscious*, CW9.1, § 284.

178   Phase IV is not only symbolically associated with the psychological basis and foundations that are all traditional meanings attached to Cancer and the fourth house. Astrological treaties *also* associate it with "the end of things." Rudhyar explains that this end, unlike that symbolized by the twelfth house, is a total end, an end without subsequent beginning. It is the consequence of defeat in the encountering in twelfth house with the

ghosts and shadows of the cycle which ends: "Then the new cycle is not a real rebirth but a descent into the abyss of final and total disintegration. There is a failure at the crucial moment of transformation and descends by progressive stages through first, second and third houses and reaches bottom, the ultimate end in the fourth." In Rudhyar, *The Astrological Houses*, ibid.

179   Stephen Hawking & Roger Penrose, *The Nature of Space and Time* (Princeton, NJ: Princeton University Press, 1996).

180   See note n°131.

181   Samy Maroun & Carlo Rovelli, "Dark Energy," in *Samy Maroun Center for Quantum Physics*, 2. Available online at http://smc-quantum-physics.com/pdf/DarkEnergy.pdf

182   That is to say, their superposition is destroyed.

183   Maroun & Rovelli, ibid., 5.

184   The Landauer principle states when the universe locally loses a data bit, a heating of the environment occurs that increases entropy of k log 2, corresponding to an energy dissipation kT log 2 where k is Boltzman constant and T the temperature.

185   Hameroff & Penrose, ibid., 571.

12. The Universe and its Fourfold Reflection

Cosmic inflation, matter/light decoupling, reflective consciousness, and beginning/end, are the four emblematic events that express, through their mirror reflections, the refining work of modelization and understanding of the universe.

The previously conjectured enantiodromic relationships have only vaguely revealed some fundamental 'Heraclitean' compensative feed-back between separate cosmological events. These limited results are probably due to the enormous backwardness in the knowledge of the psyche with respect to the study of matter. Despite the development of analytical psychology and contemporary attempts to account for consciousness, a physics of the soul has not yet caught up. Yet, Pauli's ideal of a 'background physics' is a promising breakthrough of a new way to scientifically describe and clarify these hypothetical regulatory mechanisms. Following his inspiring idea, we will now concentrate on the four cosmological events of the I-VII & IV-X polarities and confront their archetypal background to the concept of projection through both psychological and ontological perspectives.

12.1. Four Types of Psychological Projection

The four emblematic events of the universe (i.e., inflation, matter/light decoupling, reflective consciousness, and its asymptotic end/beginning) echo the fourfold diagram designed by Marie-Louise von Franz at the end of her book *Projection and Re-Collection in Jungian Psychology.*[186] It is reproduced below in figure 55. For convenience, the two poles of the horizontal axis (Matter - Objective Psyche) have been reversed with respect to the original scheme.

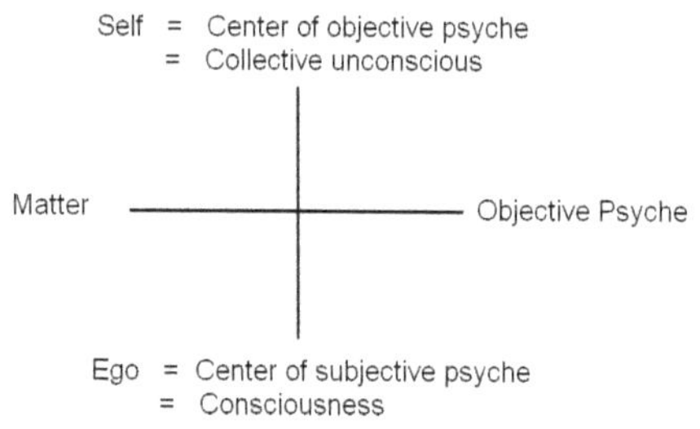

Self = Center of objective psyche
      = Collective unconscious

Matter ——————————————— Objective Psyche

Ego = Center of subjective psyche
     = Consciousness

Figure 55. The fourfold mirror reflection.
(Reprinted with permission from Open Court Publishing Company)

Psychological projections are aspects of our own inner reality of which we are unconscious and are therefore constantly projecting onto the outer world. Although nothing measurable is cast out of the psyche, we can use the metaphor of psychic energy flowing from us as from a projector onto a screen; hence, as Jung put it, an illusory relation to our environment: "projections change the world into the replica of one's own unknown face."[187] As far as the unconscious aspects of oneself – both good and bad – are acknowledged, the 'shadow' can be assimilated into the conscious personality. This is what Jung calls the individuation process, the one through which the unity of Self is achieved. Von Franz identifies four types of psychological projections, which she views as two self-referencing processes or mirror-image relations that she calls 'mirrorings.' Matter-psyche 'mirrorings' have largely been explored by Jung and Pauli through their insights on the origins of scientific theories and synchronistic phenomena. Ego-Self 'mirrorings' appear in dreams as Edinger recalls: "the Self becomes visible to the ego by being seen in dreams. Presumably the reverse is also true. Perhaps the life dramas of the ego are the dreams of the Self. Shakespeare says, 'We are such stuff as dreams are made on.'"[188]

12.2. Ego-Self Mirrorings

12.2.1. The mirroring of the ego by the Self

Dreams express something that the ego does not know or does not understand. They assist in the process of individuation. The ego consciousness in time and space is 'objectively' reflected by the eternal and infinite Self holding up a mirror to the ego.

12.2.2. The mirroring of the Self by the ego

The Self can be brought from its state of potentiality and realize itself through mirroring into the clarity of ego consciousness. Jung recounts two impressive dreams he had:

- one about a flying saucer. Usually, unidentified flying objects are seen as projection of our unconscious. At a particular time of the dream, he found himself the object of scrutiny. The UFO was pointing straight at him like a telescope. He thought: "now it turns out that we are their projections. I am projected by the magic lantern as C. G. Jung. But who manipulates the apparatus?"[189]

- another one about a yogi with his own face in deep meditation dreaming him into existence. He was frightened and awoke with the thought: "'Aha, so he is the one who is meditating me. He has a dream, and I am it.' I knew that when he awakened I would no longer be."[190]

12.3. Matter-Psyche Mirrorings

12.3.1. The mirroring of matter by the collective unconscious

Matter reflected by the psyche has already been mentioned through the investigation by Pauli on Kepler's work: we unconsciously adapt the stimuli of the external world according the archetypes of our psyche. Theories and explanations of 'material phenomena' are nothing but mirrorings of 'matter' into the human mind: initially, the physicist develops hypothetical images that he makes clear through a long process of discrimination and ordering, culminating in laws expressed in mathematical language. The latter are therefore projections of objective psyche or collective unconscious onto unknown matter. Therefore, reality can no longer be considered independently of any observer, and the object of research as Werner Heisenberg puts it "is no longer nature in itself but rather nature exposed to man's questioning, and to this extent man here also meets himself."[191]

12.3.2. The mirroring of the collective unconscious by matter

The psyche reflected by matter has also been mentioned through the synchronistic phenomena and the alchemical process. In the latter, alchemists unconsciously projected their own inner states onto external objects "Everything unknown and empty is filled with psychological projection; it is as if the investigator's own psychic background were mirrored in the darkness. What he sees in matter, or thinks he can see, is chiefly the data of his own unconscious, which he is projecting into it."[192] Jung tried to theorize the synchronistic phenomena: how can an event in the outside world (a scarab at a window) be linked to an inner psychic event (a golden scarab dream)? In a complementary relationship with the viewpoints of causality, he defined synchronicity as an "acausal ordering principle" of which astrology and divination techniques would be other examples.

Jung's and Pauli's studies on matter-psyche relations were simultaneously and 'symmetrically' divulged in their 1952 joint book *The Interpretation of Nature and the Psyche*. Pauli, faithful to his quest for symmetry in all fields of knowledge, insisted on the symmetrical side of projection which is introjection, the process of taking external images and reality into the inner world: "Do we lay our mental screen over reality or is it reality that forces itself on us and compels us to insight (projection or introjection)?"[193] In a letter to Marie-Louise von Franz (16 October 1951) he explains why there must *also* be a reflection of the objective psyche by matter: "A truly symmetrical relationship requires that the concept of *introjection*, in other words information from the outside world to the psyche, be accorded as much importance as the concept of projection, contents flowing from the psyche to the world."[194]

At first sight, a perfect symmetry between the process of matter reflected in the psyche and the process of the psyche reflected in matter (implying that synchronicity is a common phenomenon) is difficult to understand today with a so-called inanimate or lifeless matter. Gone are the days when, in the dark void of the unknown, the alchemists perceived the fiery sparks or scintillae of the 'light of nature.' But Jung, contemporary of the advent of quantum physics, regarded alchemy as a precursor to modern science. He formulated the existence of 'absolute knowledge' not mediated by the sense organs and not connected with the ego. It is not a conscious knowledge as we know it, "[...] but rather a self-subsistent 'unconscious' knowledge which I would prefer to call 'absolute knowledge.' It is not cognition but, as Leibniz so excellently calls it, a 'perceiving' which consists – or to be more cautious, seems to consist of images, of subjectless 'simulacra.'"[195]

This 'unknown knowledge' or knowledge which the subject does not know it knows, is the knowledge of the unconscious that is extended throughout all dimensions of space and time and diffused throughout the totality of existence. With information theory, physics has given new life to this hypothesis since – more than energy or matter – information has become the basic building block of all things. A quantum bit (qubit) is the smallest unit of quantum information which is not limited to two states (like, for example, in the *on* or *off* position). It can exist in a superposition of multiple states at the same time. Atoms, electrons and elementary particles have a computational power. They are quantum bits just like everything in the universe, which is supposed to have emerged by making use of its own computational capabilities.

Orch OR theory may bring a new understanding of what Jung meant by 'absolute knowledge.' As microtubule quantum events in the brain correlate with fluctuations in the structure of space–time geometry, there may exist precursors of some kind of proto-consciousness embedded in the most primitive Planck-level ingredients of space-time geometry at the unimaginably small scale of 10^{-35} m and 10^{-43} s. Like 'glittering scintillae' at a distance and a time some 20 orders of magnitude smaller than those of normal particle-physics scales and processes, they await capture at significantly larger biological scales and processes where they may be relayed and amplified to consciousness via the neurons.[196]

This 'protoconscious' information encoded in space-time geometry is reminiscent of the *lumen naturae*, the light concealed in matter and the forces of nature. Marie-Louise von Franz gives an acute summary of how Jung perceived the matter-psyche mirroring:

> Jung's observation that the reconstruction of psychic processes in the microphysical world probably occurs as continuously as the psyche perceives the external world is to be understood in the sense that this mirror-relation exist continuously in the deeper layers of the unconscious but that we become aware of it only in certain exceptional situations in which synchronistic phenomena become observable. That would mean that in the deepest layer of

the unconscious the psyche 'knows' itself in the mirror of the cosmic world and that matter 'knows' itself in the mirror of the objective psyche, but this 'knowledge' is 'absolute' in the sense that, for our ego, it is almost completely consciousness-transcending.[197]

Are there any similar 'mirrorings' in the cosmological description? We had already shown that the evolving universe matches the alchemical stages nigredo, albedo, citrinitas, and rubedo, which are just another projective process of a cosmic and spiritual drama. Let us now concentrate on the four emblematic events of the universe and their possible relation with the above fourfold mirror reflection.

12.4. The Four Emblematic Events of the Universe as Four Psychophysical Projections

The four emblematic cosmological events may be symbolized by the four moments of the Sun's yearly journey in terms of the interaction of the two universal forces in opposite polarities: 'individual' Day-Yang and 'collective' Night-Yin (Fig. 56). While matter-psyche mirroring corresponds to the I-VII axis, the ego-self mirroring corresponds to the IV-X axis.

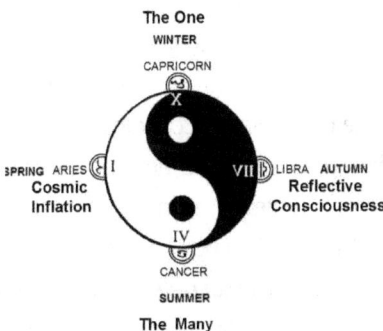

The One
WINTER
CAPRICORN
X

SPRING ARIES
I
Cosmic
Inflation

VII LIBRA AUTUMN
Reflective
Consciousness

IV
CANCER
SUMMER
The Many

Figure 56.
The 4 cosmological principles

12.4.1. Reflection of psyche by matter (or pole VII by pole I)

Cosmic inflation coincides with the Zodiacal Sign of Aries. This is the spring equinox when the Day-force, which has increased in strength while the Night-force decreased, equals in power the Night-force.

This cosmological event is the point of inception of 'matter' as an energy in the form of excitations of quantum fields. Matter evolves out of the vacuum under the form of mass-carrying particles: fermions, which consist of quarks and leptons. As a hot, dense plasma of charged particles and photons, the universe can also be considered as an information processing system, emerging as a 'Bit Bang' with qubits expressing the laws of physics – that will be subsequently reflected in relation to the ego-consciousness of an observer at fall equinox. But prior to the light-matter decoupling of the summer solstice (and emergence of multiplicities of cosmic forms), the computational processes are reflected in their 'material' counterpart.

12.4.2. Reflection of matter by psyche (or pole I by pole VII)

Reflective consciousness coincides with the zodiacal sign Libra. This is the fall equinox when the two forces are again equal, the Night-force having grown stronger ever since the beginning of the summer.

It is sometimes difficult to determine what reflects what: matter → psyche, or psyche → matter? But here, after the summer solstice, the light-matter decoupling event and the ensuing local complexification of matter allows for a negentropic evolution which will give rise to the explosive surge of reflective consciousness. For the ego-consciousness *of an 'observer-member,'* inner and outer events may be perceived as if the outer 'material' event reflects the inner psyche. Such a 'material' event would then be considered as synchronistic phenomena. But *at the level of universal reflective consciousness*, the reflective consciousness is itself a mirror. It is part of the objective psyche which reflects matter and makes knowledge possible.

The universe has become conscious of itself, and this self-knowledge persists beyond the post-material phases of its unfolding.

12.4.3. Reflection of Self by Ego (or pole X by pole IV)

The birth of galaxies and other cosmic structures coincides with the summer solstice and the Cancer sign. The Day-force reaches a maximum energy, the Night-force its lowest ebb.

A slight amount of asymmetry is a condition for the origin of the universe, and nothing could exist in a perfectly symmetrical state. It is the slight discrepancy between the amounts of matter and antimatter that enables the material universe to come into existence. The dance engaged between symmetry and asymmetry pervades the whole cyclic interplay of the One and the Many. What exists, therefore, appears as a flaw in an ideal crystalline perfection. "The universe is but a flaw in the purity of non-being."[198] It lives on the razor's edge, "at the temperature of its own destruction."[199] It narrowly escapes heat death, the point of maximum entropy, at the summer solstice, when protons and electrons manage to join, forming hydrogen atoms.

At that 'solstitial' point, light is set free. Regions of different densities are allowed to grow thanks to a precise value of the ratio of the energy needed to break up and disperse the biggest structures in our universe to their 'rest-mass energy:'[200] one part in a hundred thousand. It appears as a fine-tuned value that allows clusters and superclusters (i.e., 'proto-egos') to actualize. Matter-Light decoupling, therefore, echoes the myth of separation of Father Sky and Mother Earth. Children born between them are the seeds of consciousness-to-be, ego-like complex structures partly constituted by the inherited matter and partly by gathering disparate elements. In order to reach the Self, they must first build a strong ego and become stars: "Only a unified personality can experience life, not that personality which is split up into partial aspects, that bundle of odds and ends which also calls itself 'man.'"[201]

The constellations have been the first objects to fascinate people in the night sky, with their ordered patterns that reflected the need for order in all matters of human experience on the earth-surface. The patterns only have meaning when viewed from Earth, since the stars within a constellation may lie light-years apart. The two-dimensional images have been fully demystified by modern astronomy. In 1965, following the prediction by Gamow of a remnant heat from the Big Bang, telescopes focused in the dark space between stars and found that our sky had retained a structural memory of the time of matter/light decoupling. Over billions of years, the cosmic microwave background (CMB), left over from the time of recombination, reflected the image of the many – the seeds of galaxies today measured as slight temperature variations known as anisotropies. The latter were found with the right amplitude for structure to be accounted for by the gravitational collapse of primordial inhomogeneities, in accordance with the above fine-tuned ratio of 1 part in 100,000.

But it is the remarkable uniformity of its temperature – reflecting the quality of the One – that first brought up a puzzle. How could the size of the universe, some 380,000 years after the Big Bang, be much greater than the distance light would have traveled since the beginning? A natural explanation for the extreme smoothness in temperature has been offered by the theory of inflation. While it solves the horizon problem, it shows that the primordial seeds that grew into galaxies were planted during the first moments of the universe's existence and identify with vacuum quantum fluctuations. Quantum fluctuations or primordial seeds are like the waves of the ocean. The ocean is the quantum vacuum or what is behind the 'origin-end' of the universe. Ocean and waves are inseparable, just as the One from the Many or the Self from the Ego. And the Self-Ego axis is just another expression of the conjunction of opposites.

Primordial seeds may be awakened to (ego) consciousness – paraphrasing Hofstadter – in their *mirroring* in the unity of the fossil light. This author argues that consciousness arises as "an interaction between levels in which the top level reaches back down towards the bottom level and influences it, while at the same time being itself determined by the bottom level. In other words, a self-reinforcing 'resonance' between different levels ... The self comes into being at the moment it has the power to reflect itself."[202] Although the latter quotation is at the human level endowed with the usage of a language, it may well reflect the formation of the ego at the cosmological scale. The Self, center and circumference of the conscious and unconscious, is represented by the vacuum behind the 'origin-end' of the universe. It holds up the mirror of the cosmic background to the ego – center of the field of consciousness – so that the latter can see itself from a point outside ego-consciousness in the unconscious realm of the all-encompassing horizon.

12.4.4. Reflection of Ego by Self (or pole IV by pole X)
The beginning and the end of the cosmological cycle coincides with the winter solstice and the Capricorn sign. The Day-force is at its weakest and the Night-force at its strongest level.

At this critical point, the infinite temporal past and future of the universe come together. The heat death of the universe, implied by the second law of thermodynamics, was first conceptualized. Following the profusion of the multiplicity of the world, this unity of a state of eternal rest was relatively easier to explain than the reverse task of building the Many from the One, echoing the same difficulty in philosophy. René Thom, the inventor of catastrophe theory thought that, "at any rate, the only way we can attain an explanation, a generation of empirical diversity from one unique principle is by resorting to a process of emanation, of 'procession,' as the Neoplatonists saw very well."[203]

This first principle – the One – is beyond Being. It is described by the late Neoplatonist Proclus as a transcendent and imparticipable *cause* that pre-exists all beings and Gods that produce beings, "a cause ineffable indeed by all language, and unknown by all knowledge and incomprehensible, unfolding all things into light from itself, subsisting ineffably prior to, and converting all things to itself, but existing as the best end of all things."[204] For Proclus, the One is ineffable and should be honored in silence. But on the other hand, he states that all multiplicity participates in the One and that there is the possibility of understanding the intermediate levels between the One and our world through intellection and rational exploration. As we saw earlier, he himself had a broad interest in natural science and, beyond his rejection of the precession of the equinoxes, he had a very good knowledge of astronomy. He made a very beautiful praise of the universe, inviting us to understand it as a product of the divine intellect. However, Proclus imposes a limitation onto the human creature. The One is beyond the scope of our knowledge and cannot be accounted for by rational processes. It is better

> [...] to rest content with the negations, and by means of these to exhibit the transcendent superiority of the One – that it is neither intelligible nor intellectual, nor anything else of these things which are cognizable by us by means of our individual mental activities; for inasmuch as it is the cause of all things, it is no one of all things; and it is not the case that it is unknowable to us while being knowable to itself; for if it is absolutely unknowable to us, we do not even know this, that it is knowable to itself, but of even this we are ignorant [...] [205]

As the human mind has progressively captured nature's most complex structures, physicists are becoming aware that, approaching the Planck conditions of physics, they may be reaching a limit to understanding. Ever-new models, like fishermen's nets, are launched toward the abyss of reality, but with the awareness that science cannot provide a complete explanation of the universe. However, this pursuit reveals a desire towards the One similar to the great scientific achievements that followed the Renaissance. The unified theory of celestial and earthly phenomena, as well as the concept of unified forces of electricity and magnetism, had been fully proposed by Newton and Faraday within a unity with a God. Yet, today's search for ever-more theoretical unification in cosmology appears more as an unacknowledged yearning towards the unity of the Self.

The widely used modern technique of holography, based on the interaction of waves, appears as a 'scientific analogy' for the 'emanation' or 'projection' of the unity of the beginning-end of the universe onto the multiplicity of the post-decoupling cosmic seeds. A hologram is a three-dimensional image encoded onto a two-dimensional surface. In a recent holographic principle inspired by black hole thermodynamics – a similarity between the initial and final singularities of the universe and the event horizon of a certain type of black hole – the information that makes up the 3D reality of a region would be encoded onto the surface of the cosmological horizon of the universe. In another sense, the implicate order of David Bohm (nested in a super-implicate order equivalent to *unus mundus*) is also a holographic model. What is projected or unfolded in any type of holographic model remains unknown, although an obvious candidate could be gravity as an emergent property of information. The quantum vacuum could be the entity that represents the Self at the beginning-end of the universe as the implicated or enfolded order, while the seeds of growing cosmic structures would be the explicate or 'ego-unfolded' order.

Science is silent about explaining the One and the Self, but nothing prevents the gods grounded in the One to reveal their true nature by means of a new myth or through the evolution of present myths. Throughout his life, Jung wondered about the cosmogonic meaning of human consciousness as to the evolution of divinity and religions. In a letter to Erich Neumann in March 1959, he writes:

> Without the reflecting consciousness of man, the world is of a gigantic meaninglessness. [...] Since a creation without the reflective consciousness of man has no recognizable meaning, with the hypothesis of latent sentience a cosmogonic significance is extended to man, a true *raison d'être*. If, on the other hand, the latent meaning is attributed to the creator as a conscious plan of creation, then the question arises: why should the creator contrive this whole world phenomenon as he already knows what he could be reflected in and why should he reflect himself since he already knows what he could be reflected in and why should he reflect himself since he is already conscious of himself? To what end should he create a second, inferior consciousness alongside his omniscience? In a sense, billions of dull little mirrors of which he knows in advance what the picture will be like that they will reflect back?[206]

When Jung speaks of God, it is never about God in itself. He means the image of God in our experience, which coincides with a representation of the Self. Drawing from gnostic mythical schemas, Jung rejects a wholly transcendent Christian divinity and consider that God has both a good and a bad side. He writes:

> Man's task is ... to become conscious of the contents that press upward from the unconscious. Neither should he persist in his unconsciousness nor remain identical with the unconscious elements of his being, thus evading his destiny, which is to create more and more consciousness. As far as we can discern, the sole purpose of human existence is to kindle a light in the darkness of mere being. It may even be assumed that just as the unconscious affects us, so the increase in our consciousness affects the unconscious.[207]

Being, in some way unconscious, God (the archetypal image of God – i.e., the Self), is produced by man (i.e., the Ego) through projection, and man acts as a mirror: "Man is the mirror which God holds up before him, or the sense organ with which he apprehends his being."[208] In some way, the Self also undergoes transformation and realizes its wholeness through the Ego. Edward Edinger takes Jung seriously and expresses what would be the emerging 'new myth' of humanity:

> The new myth postulates that the created universe and its most exquisite flower, man, make up a vast enterprise for the creation of consciousness; that each individual is a unique experiment in that process; and that the sum total of consciousness created by each individual in his lifetime is deposited as a permanent addition in the collective treasury of the archetypal psyche.[209]

In some way, just as the Assumption of Mary in 1950 was an acknowledgement of a reintegration of the 'missing fourth,' the Christian myth is undergoing further gradual transformations that bring to consciousness both God and Man (the Self and the Ego) in equal need of each other in order to fulfil their wholeness. This is exemplified by a recent papal encyclical from 7 June 2017: "Jesus Christ's Gospel reveals to us that God cannot be without us: He will never be a God 'without man;' it is He who cannot be without us, and this is a great mystery! God cannot be God without man: this is a great mystery!"[210]

A similar evolution may be seen in Teilhard de Chardin's evolutionary cosmology, where God is somehow implanted from the beginning in an imperfect multiplicity. Through a natural process, information generated in the cosmos culminates in the Omega Point and actualizes the divine being: "It is Christ, in very truth, who saves – but should we not immediately add that, at the same time, it is Christ who is saved by Evolution?"[211] Teilhard's Omega Point was taken up by physicist Frank Tipler who made a quasi-religious – old Christian style – interpretation of the final state of the universe as 'omniscient.' His modelization of the future of the universe is, however, an interesting one, likely to be refuted and hopefully superseded by a more balanced model which would reflect a mutual and complementary process of transformation between the cosmic seeds and the 'unconscious' or 'supraconscious' final singularity.

Physics' difficulty with disentangling the origin-end is inherent to the fact that the physicist's mind is part of the universe. How can the eye see itself? Even if initial conditions could be explained, how about the laws of physics themselves and 'what breathed fire' into them at time zero? Like on the horizontal polarity where the psyche reflects matter just as matter reflects the psyche, the same kind of mirroring is expected on the vertical axis between Ego and Self, but with the difference that the latter psychological instances dwell at different levels of reality as expressed by what Jung called the transcendent function. This difference persists in their reflections on the phenomenal plane: the formation of cosmic seeds in the post-decoupling universe dropped below about 3,000K is relatively well understood by classical physics, while the description of the origin-end of the universe points to a domain rooted in quantum realms. As the Ego emerges from the Self and returns to the Self in the evolutive process of becoming, both instances may be represented on a circular or spiral path. In terms of consciousness, the odyssey begins with the unconscious and returns to the unconscious; the latter, following 'processing' by consciousness may be called 'superconscious.' But as Jung observes, the Self which embraces the totality of consciousness and the unconscious is also found in the center from which it orders the circumambulation of archetypes. Again, a better representation would be that of a sphere, where the circle on the equatorial plane would represent the individuation process (Fig. 57). The Ego-Self axis would *also* be found on the vertical side, with the center of the sphere being permanently occupied by the transcendental Self, heir to the Neoplatonist ineffable and imparticipable One.

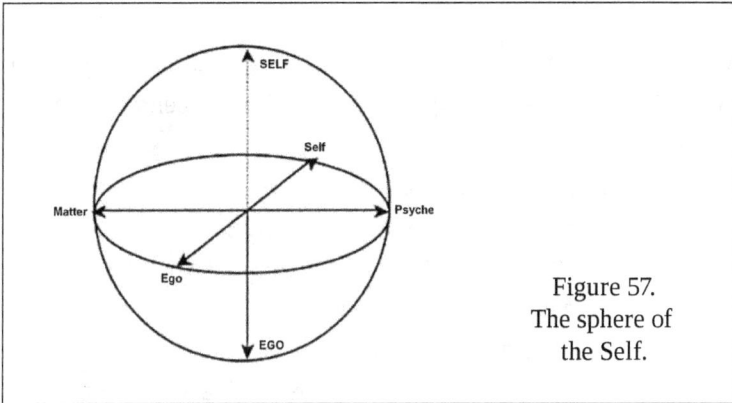

Figure 57.
The sphere of
the Self.

12.5. The Chiseling Work of the Universe

The progress of scientific knowledge and understanding of the universe improves over time with the development of theories that are tested against reality and, if refuted, new models are developed and resubmitted to observations. This process of falsification of models likely to be superseded by ones capable of giving a more comprehensive explanation of observations is equivalent to the process of the withdrawal of projections. In this way, contemporary cosmology continues to get rid of projective animism and constantly launches its theoretical nets. As strange as it may be, the prevailing evolutionary model of the universe extended by inflation and other speculative models of consciousness – but not so strange insofar as it is a projection of the transcendental *unus mundus* – shows that the universe is itself in the process of retaking its projections by remaining at the center of the fourfold mirror relation described by Marie-Louise von Franz. The four main cosmological events that stand out from its evolutionary process (inflation, matter/light decoupling, reflective consciousness and beginning/end) express through their mirror reflections the refining work of modelization and understanding of the universe. The two fundamental polarities Psyche/Matter and One/Many (Self/Ego) reflect the psychophysical background. The contrasting opposites, linked by enantiodromic processes, act as compensative mechanisms not (yet) brought to light by science, only suspected by the incredible fine-tuning of the physical constants that enable the stability of the whole system of the universe. Their consideration may open new windows likely to shed light on the 'dark enigma' and the enormous discrepancy between the value of vacuum energy predicted by quantum theory that is 120 orders of magnitude larger than what is needed to cause the observed acceleration in the universe's expansion. Using the analogy of the unworked stone made by the first Neoplatonist, Plotinus, to reveal the attainability of beauty through sculpting one's own statue, one can wonder whether the universe is not, in some way, chiseling itself. And whether or not this chiseling, this eternal process of removing the inessential, would be one and the same with the theorization process that strip off the old modelizations which hinder the representation and darken the vision. This would mean that both sculptor and sculpted – united in the art of sculpture – try, in Marie-Louise von Franz words, to "keep the inner mirror free of dust" and live in the creative current or stream of the Self – and indeed, become a part of this stream. Beauty, according to Plotinus and following Plato, has reached an infinite hidden in the splendor of the One. It is the beauty of the Intellect which confer a spiritual relationship with the One:

> If you do not yet see your own beauty, do as the sculptor does with a statue which must become beautiful: he removes one part, scrapes another, makes one area smooth, and cleans the other, until he causes the beautiful face in

the statue to appear. In the same way, you too must remove everything that is superfluous, straighten that which is crooked, and purify all that is dark until you make it brilliant. Never stop sculpting your own statue, until the divine splendor of virtue shines in you ... If you have become this ... and have nothing alien inside you mixed with yourself ... when you see that you have become this ... concentrate your gaze and see. For it is only an eye such as this that can look on the great Beauty.[212]

ENDNOTES

186   M.-L. von Franz, *Projection and Re-collection in Jungian Psychology: Reflections of the Soul*, trans. William H. Kennedy (La Salle, Il: Open Court, 1980), 187.

187   Jung, *Aion*, § 17.

188   Edward F. Edinger, *The Creation of Consciousness Jung's Myth for Modern Man* (Toronto, Canada: Inner City Books, 1984), 56, who quotes William Shakespeare, *The Tempest*, act 4, scene 1, line 156.

189   Jung, *Memories, Dreams, Reflections* (NewYork, NY: Pantheon Books, 1963), 323.

190   Ibid.

191   Werner Heisenberg, "The Representation of Nature in Contemporary Physics," in R. May (ed.), *Symbolism in Religion and Literature* (New York, NY: George Braziller, 1961).

192   Jung, *Psychology and Alchemy*, § 332.

193   Suzanne Gieser, *The Innermost Kernel: Depth Psychology and Quantum Physics. Wolfgang Pauli's Dialogue with C.G. Jung* (Berlin: Springer-Verlag, 2005), 177.

194   Ibid.

195   Jung, *Synchronicity: An Acausal Connecting Principle*, § 931.

196   Hameroff & Penrose, ibid., 558.

197   von Franz, *Projection and Re-collection*, 194.

198   Paul Valéry, *Sketch of a Serpent* (Austin, TX: Thorp Springs Press, 1987).

199   Edgar Morin, *La Méthode, 2. La vie de la vie [The Life of Life]* (Paris: Editions du Seuil, 1985).

200   Martin J. Rees, *Just Six Numbers: The Deep Forces That Shape the Universe* (London: Weidenfeld & Nicolson, 1999), 106.

201   Jung, *Psychology and Alchemy*, § 105.

202   Douglas Hofstadter, *Gödel, Escher, Bach: An Eternal Golden Braid* (New York, NY: Basic Books, 1979), 709.

203   Peer F. Bundgaard & Frederik Stjernfelt, "René Thom's Semiotic and its Sources," in *Semiosis and Catastrophes, René Thom's Semiotic Heritage*, Wolfgang Wildgen & Per Aage Brandt (eds.) (Bern, SW: Peter Lang, 2010), 76.

204   Proclus, *The Theology of Plato Volume I, Book III*, trans. Thomas Taylor (Demosthenes Koptsis, 2016), 388.

205   Proclus, *Commentary on Plato's Parmenides, Book VI*, trans. G. Morrow & J. Dillon, (Princeton, NJ: Princeton University Press, 1987), §125.

206 Jung to Dr. Erich Neumann, March 10, 1959 in *Letters 1951-1961, Vol.2.*

207 Jung, *Memories, Dreams, Reflections*, 326.

208 Jung to Jakob Amstutz, March 28, 1953 in *Letters 1951-1961, Vol. 2*, 111-2.

209 Edinger, ibid., 23.

210 Pope Francis, "General Audience: on God's Fatherhood, the Source of Our Hope." In *Zenit: The World Seen from Rome*. Available online at https://zenit.org/articles/general-audience-on-gods-fatherhood-the-source-of-our-hope

211 Pierre Teilhard de Chardin, *The Heart of the Matter*, trans. René Hague (New York, NY: Harcourt Brace Jovanovich, 1979), 92.

212 Plotinus, *Ennead 1.6.9*. In Pierre Hadot, *Philosophy as a Way of Life: Spiritual Exercises from Socrates to Foucault*, ed. Arnold I. Davidson trans. Michael Chase (Oxford, UK: Wiley-Blackwell,1995), 100.

Conclusion

We have used the hypothesis of Jung and Pauli who, reflecting on mental and physical phenomena, were led to the idea of an unknowable *psychophysical whole*. This notion that was inherited from the medieval *unus mundus* – a kind of model of our physical and sensitive world in which we unfold over time – is also a variation of the World Soul of the ancient Greeks, which was an intermediate instance between mind and matter.

This common metaphysical source is very close to the concept of the collective unconscious, whose archetypes are irrepresentable empty structures that – having been projected onto the unknown of matter and psyche – become psychic archetypal representations of the symbolic universe. In their process of invention and creation of laws or models, scientists draw their intuitions from this same transcendent unity.

Numbers-archetypes provide the means to formulate what Pauli called a 'neutral' language with regard to the psycho-physical distinction. Indeed, the language of numbers is constitutive of laws and cosmological models, while rooted in the transcendental background from which those numbers manifest by leveraging the unity of Being into intermediate planes, and according to forms or narratives that order becoming.

We have shown that the narrative of the history of the universe, such as it has been interpreted from the laws of modern physics, affords glimpses of archetypal qualities of the first four integers. Under the assumption that these traces appear at discontinuities and transitions of energy, a symbolic traditional form, mandala-like, has emerged from the interpretation of the narrative.

This symbolic quaternity, as Jung and von Franz have clearly shown, often emerges as a representation of all the symbols of the unconscious. Furthermore, a ternary rhythm has appeared as a 'dialectical process' punctuated by three of the four forces of physics that are responsible for the formation of structures in the universe.

Finally, it is a 4 × 3 structure that emerges, in which archetypes of Three and Four, harmoniously intertwined, draw the symbolic figure of the zodiac. The 4 × 3 quadrants of the 'journey of the soul' show, while integrating it, the dilemma of the 'passage of Three to Four.'

This problematic transition between the two numbers-archetypes has been known since antiquity as indicating a blockage in the flow of thought; a foreign element that implies a return to another level of reality, towards the primordial One. In terms of scientific cosmology, one should not be surprised to find echoes of this 'fourth' associated to the Planck era events which today arouse so much research.

As for the first integers on which it is based, the zodiac wheel is a 'neutral' language that allows us to go beyond psychophysical dualism. Building on the subtly asymmetrical symmetry of the zodiac, we questioned the heuristic potential of Heraclitean enantiodromia associated with the 6×2 polarities of the main traditional 'aspects.' Events associated with initial $t = 0$ and final $t = \infty$ 'theoretical singularities' subsumed into a single pole, have thus become opposed to the 'matter/light decoupling' event, whose pale reflection reaches us today in the radiant background of the sky.

Post-decoupling events of the material era, whose physics is relatively well-known and mastered in contemporary laboratories, have appeared in polarity with less-known events of the Planck era. Thus, the possibility emerges to suggest, if not invariants, at least morphological similarities, using symmetry inherent to the reading of the grid.

Among the six polarities highlighted by the reading of the grid, one of the polarities has been pointed out in the 2000s in some promising research on the quantum origin of consciousness.

Already suggested in our 1994 book, this analogy of form between cosmic inflation and consciousness remained fragile due, first of all, to the *ad hoc* nature of the theory of cosmological inflation and, secondly, the Orch OR theory that then seemed inadequate in a room-temperature environment such as the brain.

But since then, the theory of cosmological inflation has gained more and more credibility, until it has become unavoidable, as it naturally appears in various independent primordial universe models. On the other hand, the Orch OR theory recently gained credibility with new teams dedicated to the study of the brain.

While science is irreplaceable in the process of projection withdrawal, allowing for a better understanding of reality, it should not be separated from culture. The path opened by Carl Gustav Jung and Wolfgang Ernest Pauli, especially in their rediscovery of a 'neutral' language of numbers, allows for the connecting of many areas. These number-archetypes underlie both the psyche and matter, reflecting a unitary psychophysical reality. They enrich the understanding of mathematics as an expression of symmetry and unifying power, allowing for recognition at the same level of dignity of both sides of reality constituted by the physical and the mental, the quantitative and the qualitative.

Bibliography

Abbot, Francis E. (2016). "A Study of Religion: The Name and the Thing [5th Sunday Afternoon Lectures for 1872, Horticulture Hall Boston]". In *The Pamphlet Collection of Sir Robert Stout: Volume 37, The Derivation from 'Relegere.'* Available at from http://nzetc.victoria.ac.nz/tm/scholarly/tei-Stout37-t15-body-d2-d3.html

Abellio, Raymond. (2006). In *Dictionary of Gnosis & Western Esotericism, 1–2, 1060–1061.* Ed. Wouter J. Hanegraaff et al. Leiden, Netherlands: Koninklijke Brill NV Available online at http://www.antropologias.org/files/downloads/2012/12/123.pdf

Aras. (2016). The Archive for Research in Archetypal Symbolism. Available online at https://aras.org/concordance/content/enantiodromia

Atmanspacher, Harald. (2014). "20th Century Versions of Dual-Aspect Thinking." In *Mind & Matter Vol.12(2).* Available online at http://www.mindmatter.de/journal/abstracts/mmabstracts12_2.html#atm

Atmanspacher, Harald and Christopher A. Fuchs eds. (2014). *The Pauli-Jung Conjecture.* Exeter: Imprint Academic.

Barnett, Lincoln. (1966). *The Universe and Dr. Einstein, A Clear Explanation of Einstein's Theories.* New York, NY: Harper and Row.

Brach, Jean-Pierre. (1993). "Histoire des courants ésotériques et mystiques dans l'Europe moderne et contemporaine." In École pratique des hautes études, Section des sciences religieuses / Vol. 106/ N°102 (1993): 338. Available online at http://www.persee.fr/doc/ephe_0000-0002_1993_num_106_102_14924

Bundgaard Peer F. & Frederik Stjernfelt. (2010). "René Thom's Semiotic and Its Sources." In *Semiosis and Catastrophes, René Thom's Semiotic Heritage.* Wolfgang Wildgen & Per Aage Brandt (eds.) Bern, SW: Peter Lang.

Cassé, Michel. (1993). *Du vide et de la création.* Paris, Odile Jacob.

Cazenave, Michel. (Ed.). (1984). *Science and Consciousness: Two Views of the Universe: Edited Proceedings of the France-Culture and Radio-France Colloquium,* Cordoba, Spain, Oxford, UK: Pergamon Press.

Cazenave, Michel. (1995). "Synchronicité, physique et biologie." In Cazenave et al. *La synchronicité, l'âme et la science.* Paris: Albin Michel.

Cazenave, Michel. (1996). *La science et l'âme du monde.* Paris: Albin Michel.

Cazenave, Michel. (2003). "General discussion following Marc Lachièze-Rey's" at *Third Symposium Brussels, 3-6 June 2003, Science and its Representations.* Talk recorded for the programme "L'éloge du savoir" presented by Michel Cazenave and broadcast on France Culture Radio on 4 September 2003.

Cazenave, Michel. (2005). "Les mathématiques et l'âme chez Proclus." In Cazenave et al. *De la science à la philosophie : y a-t-il une unité de la connaissance ?* Paris: Albin Michel.

Chaisson, Eric & Steve Mcmillan. (2008). *Astronomy Today 6th ed.* New York, NY: Pearson Education. Available online at http://www.as.utexas.edu/astronomy/education/fall09/scalo/secure/301F09.Ch27slides.pdf

Collins, Rodney. (1971). *The theory of Celestial Influence.* New York, NY: Samuel Weiser.

Dawson, John W. Jr. (1997). *Logical Dilemmas: The life and work of Kurt Gödel.* Wellesley, MA: A.K. Peters, Ltd.

Deutsch, David. (1997). *The Fabric of Reality.* New York, NY: Viking Adult.

Dobbs, Betty Jo Teeter. (2002).*The Janus Faces of Genius: The Role of Alchemy in Newton's Thought.* Cambridge, UK: Cambridge University Press.

Drake, Stillman. (1989). *Galileo at Work: His Scientific Biography.* Chicago, IL: University of Chicago Press.

Durand, Gilbert. (1999). *The Anthropological Structures of the Imaginary.* Trans. by Margaret Sankey and Judith Hatten. Mount Nebo, Australia: Boombana Publications.

Dyson, Freeman. (1964). "Mathematics in the physical sciences." In *Scientific American*, Vol. 211, Issue 3, September 1.

Dyson, Freeman. (1985). *The Origins of Life.* New York, NY: Cambridge University Press.

Eddington, A. S. (2012). *The Nature of the Physical World.* Cambridge, UK: Cambridge University Press.

Edinger, Edward F. (1984). *The Creation of Consciousness Jung's Myth for Modern Man.* Toronto: Inner City Books.

Edinger, Edward F. (1986). *The Bible and the Psyche: Individuation Symbolism in the Old Testament.* Toronto: Inner City Books.

Edinger, Edward F. (1995). *The Mysterium Lectures: A Journey through C.G. Jung's Mysterium Coniunctionis.* Toronto: Inner City Books.

Enz, Charles P. (2002). *No Time to be Brief: A Scientific Biography of Wolfgang Pauli.* New York, NY: Oxford University Press.

Franz, M.-L. von. (1974). *Number and time: Reflections leading toward a unification of depth psychology and physics.* Evanston, IL: Northwestern University Press.

Franz, M.-L. von. (1980). *Projection and Re-collection in Jungian Psychology: Reflections of the Soul.* Trans. by William H. Kennedy, La Salle, IL: Open Court.

Franz, M.-L. von. (1992). *Psyche and Matter.* Trans. by Michael H. Kohn, Boston & London: Shambhala.

Franz, M.-L. von.(1995). *Creation Myths.* Boston, MA: Shambhala.

Galileo Galilei. (1967). *Dialogue Concerning the Two Chief World Systems, (1632).* Trans. by Stillman Drake. Berkeley and Los Angeles: University of California Press.

Gardner, Martin. (2005). *The New Ambidextrous Universe: Symmetry and Asymmetry from Mirror Reflections to Superstrings.* Mineola, NY: Dover Publications.

Gieser, Suzanne. (2005). *The Innermost Kernel: Depth Psychology and Quantum Physics. Wolfgang Pauli's Dialogue with C.G. Jung.* Berlin: Springer-Verlag.

Granet, Marcel. (1999). *La Pensée chinoise.* Paris: Albin Michel.

Granet, Marcel. (2012). *La Pensée chinoise.* Trans. by Robert Lafleur in "Round and Square: Chinese Civilization and Thought." Available online at http://robert-lafleur.blogspot.fr/2012/04/la-pensee-cyclique-9chinese-thought-and.html.

Guénon, René. (2001). *The Reign of Quantity and The Signs of The Times.* Trans. by Lord Northbourne. Hillsdale, NY: Sophia Perennis et Universalis.

Guénon, René. (2004). *Symbols of Sacred Science.* Trans. by Henry D. Fohr. Hillsdale, NY: Sophia Perennis and Universalis.

Guinard, Patrice. *Avatars of the Astrological Zodiac.* Available online at http://cura.free.fr/25avazod.html

Hadot, Pierre. (1995). *Philosophy as a Way of Life: Spiritual Exercises from Socrates to Foucault.* Trans. by Michael Chase, ed. by Arnold I. Davidson. Oxford, UK: Wiley-Blackwell.

Halberg et al. (2006). "Chronobiology's progress I." In *Journal of Applied Biomedicine* 4: 1–38, ISSN 1214-021X. Available online at http://jab.zsf.jcu.cz//4_1/halberg.pdf

Hameroff, Stuart & Roger Penrose. (2016). "Consciousness in the Universe: An Updated Review of the 'Orch OR' Theory." In *Biophysics of Consciousness: A Foundational Approach.* Ed. R. R. Poznanski et al. Singapore: World Scientific Publishing. Available at http://www.consciousness.arizona.edu/documents/Hameroff-PenroseUpdatedReviewof OrchOR2016b2237_Ch-14_Revised-2-3.pdf

Hawking, Stephen & Roger Penrose. (1996). *The Nature of Space and Time.* Princeton: Princeton University Press.

Hawkins, Stephanie L. (2011). *William James, Gustav Fechner, and Early Psychophysics.*Front Physiol. 2: 68. Doi: 10.3389/fphys.2011.00068

Hegel, G. W. F. & Arnold V. Miller. (2004). *Hegel's Philosophy of Nature: Being Part Two of the Encyclopaedia of the Philosophical Sciences (1830).* Oxford/New York: Oxford University Press.

Hegel, G. W. F. (2010). *The Science of Logic.* George di Giovanni (ed., tr.), Book Two, Cambridge: Cambridge University Press. Available on line at https://academiaanalitica.files.wordpress.com/2016/10/georg_wilhelm_ friedrich_hegel__the_science_of_logic.pdf

Heisenberg, Werner. (1961). "The Representation of Nature in Contemporary Physics." In R. May (ed.), *Symbolism in Religion and Literature.* New York, NY: George Braziller.

Hillman, James. (2010). *Alchemical Psychology: The Uniform Edition of the Writings of James Hillman, vol.5.* Putnam, CT: Spring Publications.

Hofstadter, Douglas. (1979). *Gödel, Escher, Bach: An Eternal Golden Braid.* New York, NY: Basic Books.

Jacobi, Jolande. (1959). *Complex, archetype and symbol in the psychology of C.G. Jung .* Trans. by R. Mannheim. Princeton, NJ: Princeton University Press.

Jamsthaler, Herbrandt. (1625). *Viatorium spagyricum. Das ist: Ein gebenedeyter spagyrischer Wegweiser.* Frankfurt am Main. Doi.org/10.3931/e-rara-7106

Jardine, Nick. (2000). "Koyré's Kepler/Kepler's Koyré." In *History of Science 38: 363–376.* Available from http://adsabs.harvard.edu/ full/2000HisSc..38..363J

Jung, C.G. (1954). *The Practice of Psychotherapy.* CW 16. Princeton, NJ: Princeton University Press.

Jung, C. G. (1963). *Memories, Dreams, Reflections.* New York, NY: Pantheon Books.

Jung, C. G. and M.-L. von Franz. (1964). *Man and his symbols.* Garden City, NY: Doubleday.

Jung, C. G. (1969, 1970). *Psychology and Religion: West and East.* CW 11, 2nd ed., tr. R.F.C. Hull. Princeton, NJ: University Press (1969) ISBN 978-0-691-09759-6., Taylor Francis Ltd, (1970) ISBN 978-0-415-06606-8.

Jung, C.G. (1969, 1991). *Aion: Researches into the Phenomenology of the Self.* CW 9, Part II, 2nd ed., tr. R.F.C. Hull. Princeton NJ: University Press (1969) ISBN 978-0-691-09759-6 Taylor Francis Ltd, (1991) ISBN 10: 0415064767.

Jung, C. G.(1973). *Synchronicity: An Acausal Connecting Principle.* CW 8. 2nd ed. Princeton, N.J.: Princeton University Press.

Jung, C. G. (1975). *Letters 1906-1950, Vol. 1.* Ed. G. Adler with A. Jaffé. Princeton NJ: Princeton University Press.

Jung, C.G. (1975). *Letters 1951-1961,Vol. 2.* Ed. G. Adler with A. Jaffé. Princeton NJ: Princeton University Press.

Jung, C. G. (1977). *C.G. Jung Speaking: Interviews and Encounters.* Ed. W. Mcguire and R.F.C. Hull. Princeton, NJ: Princeton University Press.

Jung, C. G. (1980, 1989). *Psychology and Alchemy*. CW 12, tr. R.F.C. Hull. Princeton, NJ: Princeton University Press (1980) ISBN 978-0-691-01831-7. Taylor Francis Ltd, (1989) ISBN 978-0-415-03452-4.

Jung, C. G. (1990). *The Archetypes and the Collective Unconscious, CW 9.1.* Princeton, NJ: Princeton University Press.

Jung, C.G. (1996). *The Psychology of Kundalini Yoga: Notes of the Seminar Given in 1932 by C.G. Jung.* Princeton, NJ: Princeton University Press. Available online at https://monoskop.org/images/0/08/Jung_Gustav_Carl_ The_Psychology_of_Kundalini_Yoga_1932.pdf

Klibansky, R. et al. (1964). *Saturn and Melancholy: Studies in the History of Natural Philosophy, Religion, and Art.* New York NY: Basic Books.

Koestler, Arthur. (2014). *The Sleepwalkers: A History of Man's Changing Vision of the Universe.* London: Penguin Classics.

Kragh, Helge. (1999). *Cosmology and Controversy.* Princeton, NJ: Princeton University Press.

Kumar, Manjit. (2009). *Quantum: Einstein, Bohr, and the Great Debate about the Nature of Reality.* New York, NY: W.W. Norton.

Lambert, Dominique. (2015). *The Atom of the Universe; The Life and Work of Georges Lemaître.* Kraków, Poland: Copernicus Center Press.

Lamboy, Véronique. (1982). *L'art du Zodiaque.* Chambéry, France: CREER éditions.

Lao Tsu, *Lao Tzu: Tao Te Ching.*(1989). Trans. by Gia-Fu Feng and Jane English. New York, NY: Vintage Books.

Leibniz, G.W. (2012). *Discourse on Metaphysics and Other Writings.* Ed. by Peter Loptson, Peterborough, Ontario: Broadview Press.

Lemaître, Georges. (1936). "La culture catholique et les sciences positives" (séance du 10 septembre 1936) in *Actes du VIe congrès catholique de Malines*, Vol. 5, Culture intellectuelle et sens chrétien, Bruxelles, VIe Congrès Catholique de Malines.

Lennon, J., & McCartney, P. (1967). *Strawberry Fields Forever* [Recorded by The Beatles]. On Parlophone, R 5570, UK (7"/Single).

Lennon, J., & McCartney, P. (1968). "Everybody's Got Something to Hide Except Me and My Monkey" [Recorded by The Beatles]. On *"The Beatles (White Album)."* London: EMI

Leroi-Gourhan, André. (2012). *L'homme et la matière.* Paris, Albin Michel.

Lévy-Leblond, J.-M. (2008). "(Re)mettre la science en culture: de la crise épistémologique à l'exigence éthique." Available online at https://www.researchgate.net/profile/Jean-Marc_Levy-Leblond/publications

Lévy-Leblond, J.-M. (1990). "Did the big bang begin?" In *American Journal of Physics* 58. Available online at https://www.researchgate.net/profile/Jean-Marc_Levy-Leblond/publications

Lorenz, Konrad. (1943). "Die angeborenen Formen möglicher Erfahrung."
In *Zeitschrift für Tierpsychologie 5.*

Maroun, Samy & Carlo Rovelli. *Samy Maroun Center for Quantum Physics.*
Available online at http://smc-quantum-physics.com/pdf/DarkEnergy.pdf

McFarland Solomon, Hester. (2007). "The transcendent function and
Hegel's dialectical vision." In *Who Owns Jung?* Ed. by Ann Casement,
London: Karnac.

McIntosh, Christopher. (1992). *The Rose Cross and the Age of Reason.* Leiden,
Netherlands & New York: Brill.

Meier, C.A. Ed. (2001). *Atom and Archetype: The Pauli/Jung Letters, 1932-1958.*
Princeton NJ: Princeton University Press.

Meyer, François. (1985). "Temps, devenir, évolution." In *Communications 41,*
pp. 111-122. DOI : 10.3406/comm.1985.1611

Miller, Arthur. (2010). *137: Jung, Pauli, and the Pursuit of a Scientific Obsession.*
New York, NY: W.W. Norton & Company.

Mo, Houjun & Frank Van den Bosch & Simon White. (2010). *Galaxy Formation
and Evolution.* Cambridge, UK: Cambridge University Press. Available
online at https://www.astro.umd.edu/~richard/ASTRO620/MBW_Book_
Galaxy.pdf

Morin, Edgar. (1985). *La Méthode, 2. La vie de la vie.* Paris: Editions du Seuil.

Morin, Jean-Baptiste. (1974). *Astrosynthesis: The Rational System of Horoscope
Interpretation According to Morin de Villefranche.* Trans. by Lucy Little.
New York, NY: Zoltan Mason Emerald Books.

Negre, Alain. (1994). *Entre science et astrologie.* Paris: L'Harmattan.

Orsucci, Franco F. & Nicoletta Sala. (2008). *Reflexing Interfaces: The Complex
Coevolution of Information Technology.* Hershey, PA: Idea Group.

Ostrowski-Sachs, Margaret & C.G. Jung. (1977). *From conversations with
C. G. Jung.* Küsnacht, Hornweg 28, ZH: C.-G.-Jung-Institut Zürich.

Pauli, Wolfgang. (1994). *Writings on Physics and Philosophy.* Ed. by C. Enz and
K. von Meyenn. Berlin Heidelberg: Springer-Verlag.

Penrose, Roger. (2016). *Fashion, Faith, and Fantasy in the New Physics of the
Universe.* Princeton NJ: Princeton University Press.

Plato. (1989). "Timaeus." In *The Collected Dialogues of Plato.* Trans.
Benjamin Jowett, ed. E. Hamilton and H. Cairns. Princeton NJ:
Princeton University Press.

Pope Francis. (2017). "General Audience: on God's Fatherhood, the Source of
Our Hope." In *Zenit: The World Seen from Rome.* Available online at
https://zenit.org/articles/general-audience-on-gods-fatherhood-the-source-of-
our-hope

Proclus. (1987). *Commentary on Plato's Parmenides*. Trans. by Glenn R. Morrow and John M. Dillon. Princeton NJ: Princeton University Press.

Proclus. (1992). *A Commentary on the First Book of Euclid's Elements*. Trans. by Glenn R. Morrow. Princeton NJ: Princeton University Press.

Proclus. (2016). *The Theology of Plato Volume I, Book III*. Trans. by Thomas Taylor, Demosthenes Koptsis.

Raël, Leyla. (1983). *The Essential Rudhyar: An Outline and an Evocation*. Palo Alto, CA: Rudhyar Institute for Transpersonal Activity. Available online at http://www.khaldea.com/rudhyar/ess/ess_13.shtml

Rees, Martin J. (2003). *Just Six Numbers: The Deep Forces That Shape the Universe*. London: Weidenfeld & Nicolson.

Rudhyar, Dane. (1972). *The Astrological Houses, The Spectrum of Individual Experience*. Garden City, NY: Doubleday. Available online at http://khaldea.com/rudhyar/astroarticles/problemsweallface_1.php

Rudhyar, Dane. (1978). *The Pulse of Life*. Boulder & London: Shambhala. Available online at http://www.khaldea.com/rudhyar/pofl/pofl_p1p1.shtml/

Rudhyar, Dane. (1983). *Rhythm of Wholeness, A Total Affirmation of Being*. Wheaton, IL: Quest Books, Available online at http://www.khaldea.com/rudhyar/rw/rw_c4_p2.shtml

Schelling, Frederich, W. J. (2000). *The Ages of the World, Third Version (c. 1815)*. Trans. by Jason M. Wirth. Albany, NY: State University of New-York Press.

Shayegan, Daryush. (1991). *Qu'est ce qu'une révolution religieuse?* Paris: Albin Michel.

Teilhard de Chardin, Pierre. (1979). *The Heart of the Matter*. Trans. René Hague, New York, NY: Harcourt Brace Jovanovich.

Theon of Smyrna. (1979). *Mathematics Useful for Understanding Plato*. Trans. from the 1892 Greek/French edition of J. Dupuis by Robert & Deborah Lawlor. San Diego, CA: Wizards Bookshelf.

Tipler, Frank. (1994). *The Physics of Immortality: Modern Cosmology, God and the Resurrection of the Dead*. New York, NY: Doubleday & Co.

Tipler, Frank. (2005). *The Structure of the World from Pure Number*. Rep. Prog. Phys. 68. 897–964 doi:10.1088/0034-4885/68/4/R04.

Tipler, Frank. (2007). *The Physics of Christianity*. New York, NY: Doubleday & Co.

Trubetskova, Irina. (2004). *Vladimir Ivanovich Vernadsky and his Revolutionary Theory of the Biosphere and the Noosphere*. Available online at http://www-ssg.sr.unh.edu/preceptorial/Summaries_2004/Vernadsky_Pap_ITru.html

Tuszynski, Jack (ed.) (2006). *The Emerging Physics of Consciousness.* Berlin, Heidelberg and NewYork: Springer.

Valéry, Paul. (1987). *Sketch of a Serpent (Ébauche d'un serpent).* Austin, TX: Thorp Springs Press.

Vinogradoff, Michel. (2010). *L'Esprit de l'Aiguille – l'apport du Yi Jing à la pratique de l'acupuncture.* Paris: Springer-Verlag.

Vitiello, Giuseppe. (2015). *The Aesthetic Experience as a Characteristic Feature of Brain Dynamics.* Florence: Firenze University Press. Available online at: http://dx.doi.org/10.13128/Aisthesis-16207 .

Voelke, André-Jean. (1990). "Vide et non-être chez Leucippe et Démocrite." In *Revue de théologie et de philosophie* 122. Available online at Doi. org/10.5169/seals-381417

Welchman Alistair & Judith Norman.(2010). "Creating the Past: Schelling's Ages of the World." In *Journal of the Philosophy of History, Vol. 4, Issue 1.* DOI: 10.1163/187226310X490034

Welchman, Alistair. (2014). "Schelling's Moral Argument for a Metaphysics of Contingency." In *Nature and Realism in Schelling's Philosophy.* Ed. by Emilio Carlo Corriero and Andrea Dezi. Turin, Italy: Accademia University Press

Weyl, Hermann. (2009). *Philosophy of Mathematics and Natural Science.* Trans. by Olaf Helmer, Princeton, NJ: Princeton University Press.

Wheeler, John, A. (1984). "This Participatory Universe." In *A Passion to Know: Twenty Profiles in Science.* Ed. Allen L. Hammond. New York, NY: Scribners.

Zizzi, Paola. (2000). *Emergent Consciousness: From the Early Universe to our Mind.* Available online at https://arxiv.org/abs/gr-qc/0007006

Zizzi, Paola.(2007). *The 'Big Wow' Theory.* Available online at http://www.quantumbionet.org/eng/index.php?pagina=60.

www.ingramcontent.com/pod-product-compliance
Lightning Source LLC
Chambersburg PA
CBHW070911270326
41927CB00011B/2526